FIAS Interdisciplinary Science Series

More information about this series at http://www.springer.com/series/10781

Peter O. Hess · Mirko Schäfer
Walter Greiner

Pseudo-Complex General Relativity

 Springer

Peter O. Hess
Instituto de Ciencias Nucleares
Universidad Nacional Autónoma de México
Mexico City, Distrito Federal
Mexico

Walter Greiner
Frankfurt Institute for Advanced Studies
University of Frankfurt
Frankfurt am Main, Hessen
Germany

Mirko Schäfer
Frankfurt Institute for Advanced Studies
University of Frankfurt
Frankfurt am Main, Hessen
Germany

FIAS Interdisciplinary Science Series
ISBN 978-3-319-25060-1   ISBN 978-3-319-25061-8   (eBook)
DOI 10.1007/978-3-319-25061-8

Library of Congress Control Number: 2015952035

Springer Cham Heidelberg New York Dordrecht London
© Springer International Publishing Switzerland 2016

Printed on acid-free paper

Springer International Publishing AG Switzerland is part of Springer Science+Business Media
(www.springer.com)

Preface

General relativity (GR), one of the most and best checked physical theories of our time, exhibits singularities: The theory predicts that when a sufficient large mass collapses, no known force is able to stop it until all mass is concentrated at a point. The theory also predicts so-called *coordinate singularities*. These are singularities in the metric which vanish after a transformation to different coordinates. For example, when an astronaut falls freely towards a black hole, he will not see anything special, except the gravitational force with its deadly tidal effect. However, a fixed observer at a safe distance will see at a certain distance from the center, the *Schwarzschild radius* for a nonrotating black hole, an event horizon. *No informa-tion can reach the observer from places at smaller radial distances*! This is a rather discomforting observation, telling that part of the space is excluded from the observation by a nearby observer. On the other hand: Why should GR still be valid in extremely strong gravitational fields, as one encounters near the Schwarzschild radius?

This was the reason why two of the authors of this book (P.O. Hess and W. Greiner) started to discuss this point several years ago. We believe that *no acceptable physical theory should have a singularity (!)*, not even a coordinate singularity of the type discussed above! The appearance of a singularity shows the limitations of the theory. In GR this limitation is the strong gravitational force acting near and at a supermassive concentration of a central mass. There are other very successful theories, like the Quantum Electrodynamics (QED), which exhibits singularities, infinities, due to taking into account the very large momenta corre-sponding to very small distances in space-time. Most of the physicists would agree that any field theory should not apply at very small distances. Methods of regu-larizing field theories have been developed, giving a recipe how to remove the infinite contribution. But that is what they are: *Recipes*! In 2007 the authors of this book published a new field theory, called *pseudo-complex Field Theory*, where they introduced pseudo-complex variables, which will play an important role in this book. Owing to the extension to pseudo-complex fields and operators, it is shown that the theory is automatically regularized. This is due to the appearance of a minimal length as a *parameter*. Because it appears only as a parameter, Lorentz transformation does not affect it, thus, *all continuous and discrete symmetries of nature are maintained*! However, due to the extremely small effects and the minimal length, there is no hope to measure the deviations in near future.

This was the reason why we started to look for extreme physical situations, such as strong gravitational fields near a large mass. The first question was: Is there a possibility to avoid the formation of the event horizon? This would mean that the large mass concentrations, for example at the center of galaxies, are still there but these objects are *no black holes*! It will be shown that there is one natural algebraic extension of GR, namely to pseudo-complex (pc) coordinates. We developed the *pseudo-complex General Relativity* (pc-GR) and found several observational effects which can be measured in near future (See Chap. 5 of this book). The *very long baseline interferometry* (VLBI), to which ALMA, the European observatory in the Atacama dessert in Chile belongs, will be able to resolve the central massive objects at the centers of our galaxy and in M87. Thus, as GR, also pc-GR is a testable theory!

This book contains several exercises with explicit and detailed solutions. It is therefore also of interest for students working in GR. Many of the exercises correspond to considerations not published in text books or at least not in detailed form. We therefore are convinced that this book is helpful also for students only starting to work in General Relativity.

The book is divided into seven chapters. In Chap. 1, the necessary basis is led to deal with pc-variables. This chapter is necessary to understand the content from the second chapter and further on. The noninterested reader can skip it but surely he will have to return soon to the first chapter.

Chapter 2 is a central piece of this book, where the pc-GR is introduced and the basic philosophy is discussed. First, a historical overview is given on former attempts to extend GR (which includes Einstein himself), all with distinct motivations. It will be shown that the only possible algebraic extension is to introduce pc-coordinates, otherwise for weak gravitational fields, nonphysical ghost solutions appear. *Thus, the need to use pc-variables.* We will see that the theory contains a minimal length with important consequences. After that, the pc-GR is formulated and compared to the former attempts. A new variational principle is introduced, which requires in the Einstein equations an additional contribution. Alternatively, the standard variational principle can be applied, but one has to introduce a constraint with the same former results. The additional contribution will be associated to vacuum fluctuation, whose dependence on the radial distance can be approximately obtained, using semiclassical quantum mechanics. *The main point is that pc-GR predicts that mass not only curves the space but also changes the vacuum structure of the space itself.* In the following chapters, the minimal length will be set to zero, due to its smallness. Nevertheless, the pc-GR will keep a remnant of the pc-description, namely that the appearance of a term, which we may call "dark energy," is inevitable.

The first application will be discussed in Chap. 3, namely solutions of central mass distributions. For a nonrotating massive object, it is the pc-Schwarzschild solution; for a rotating massive object, the pc-Kerr solution; and for a charged massive object, it will be the Reissner–Nordström solution. This chapter serves to become familiar on how to resolve problems in pc-GR and on how to interpret the results. One of the main consequences is that we can eliminate the event horizon and thus, *there will be no black holes*! The huge massive objects in the center of nearly any galaxy and the so-called galactic black holes are within pc-GR still there, but with the absence of an event horizon!

Chapter 4 gives another application of the theory, namely the Robertson–Walker solution, which we use to model different outcomes of the evolution of the universe. New solutions will appear as the limit of constant acceleration, the limit of zero acceleration after a period of a nonzero acceleration. We also discuss the possibility of an oscillating universe, with repeated big bangs, with no need to explain the smoothness of the universe.

The success of a theory depends on the capability to predict new phenomena. Chapter 5 is just dedicated to this purpose. We will see that at a large distance from a large massive object, GR and pc-GR will show no differences. However, near the Schwarzschild radius significant deviations of pc-GR from GR are predicted. The orbital frequency of a particle in a circular orbit and stable orbits in general will be calculated. As a distinct feature, in pc-GR there will be a maximal orbital frequency. We show that above a given spin of the star, there will be no *innermost stable circular orbit* (ISCO) and an accretion disk will reach the surface of the star. This has important consequences for the physics of the accretion disk: It will appear brighter (emit more light) and due to the maximum in the orbital frequency, a dark ring is predicted by pc-GR. Also the redshift will be calculated. This is of great importance: One observes so-called *quasi-periodic oscillations* (QPO) and the redshift of Fe Kα lines. Knowing the orbital frequency of a QPO and the redshift, GR and pc-GR get for each observable a radius for the position of the QPO. Both radii, obtained from both observables, should coincide. They do not in GR, but they do in pc-GR! Of course, this depends still on the interpretation of the nature of the QPO and the discussion is still on.

In Chap. 6, neutron stars are discussed and a primitive model for the coupling of mass to the dark energy is proposed. This chapter is of conceptional nature and is meant to show that large masses for neutron stars can be obtained. The jewel of this chapter is the discussion of the so-called *energy conditions*. They are used to see if an ansatz for an energy–momentum tensor, treating for example ideal fluids, makes sense. We found no book or article in the literature where these conditions are treated as extensively as here with detailed solutions. Thus, this chapter serves also for people interested only in the standard theory of GR.

Finally, in Chap. 7, the geometric differential structure of pc-GR is investigated. The motivation for this chapter is to complete the presentation of pc-GR in a rigorous manner. For a noninterested reader of differential geometry, this chapter can be skipped. However, he may find it to be useful, to learn more of this topic. No explicit knowledge of differential geometry is required, because all necessary definitions will be given. This makes this chapter especially useful.

The book appears within the series of the *FIAS Lecture Notes*, which is meant to publish on topics of interdisciplinary interest and new developments. We think that this is an ideal place for resuming all results obtained within pc-GR.

Finally, we would like to express our sincere thanks to all the people who contributed with their help to the realization of this book. We thank Gunther Caspar and Thomas Schönenbach for their contribution to Chaps. 3 and 5, Thomas Boller and Andreas Müller for their contribution to Chap. 5 and Isaac Rodríguez for his contribution to Chap. 6. The Chap. 3 is based mainly on the master theses of Thomas Schönenbach and Gunther Casper, Chap. 5 is on the Ph.D. thesis of Thomas Schönenbach, and Chap. 6 is based on the Ph.D. thesis of Isaac Rodríguez. We acknowledge useful comments by J. Kirsch. We also thank Laura Quist for their patience and logistic help. P.O.H. wants also to acknowledge financial help from DGAPA-PAPIIT (IN100315). M.S. acknowledges the support from Stiftung Polytechnische Gesellschaft.

Mexico City   Peter O. Hess
Frankfurt am Main   Mirko Schäfer
Frankfurt am Main   Walter Greiner
March 2015

Contents

List of Figures

Chapter 1
Mathematics of Pseudo-complex General Relativity

In this chapter we give a short introduction to the mathematics of pseudo-complex variables and functions. For now, we provide a rather informal presentation of the necessary mathematical tools, such that the reader is familiarized with the notation and concepts and is prepared to follow the construction of the new theory of pseudo-complex General Relativity in the next chapters. In Chap. 7 we will then treat various concepts of pseudo-complex mathematics in a more formal and extensive way.

1.1 Definitions and Properties

Every physicist is familiar with the algebra of real numbers $\mathbb{R}$ and the calculus of real valued functions. We know how to add, subtract, multiply and divide these numbers, and have a clear understanding of such terms as "0" as a zero element of addition, "1" as a unity element of multiplication, and $-x$ and x^{-1} as the inverse elements with respect to addition and multiplication of the element x, respectively. But sometimes the real numbers are just not enough to solve a certain problem, and we want to construct another, more general number system. A familiar problem is finding a solution of the equation $x^2 = -1$, which is not possible within the realm of real numbers, but has a solution using the well-known complex numbers $\mathbb{C}$. An analogy from physics is the historical approach of Dirac, who searched for a relativistic covariant wave equation of the Schrödinger form

$$i\hbar \frac{\partial \psi}{\partial t} = \hat{H} \psi, \tag{1.1}$$

which was not possible using only real or complex numbers, but needed the incorporation of the algebra of Dirac matrices.

© Springer International Publishing Switzerland 2016
P.O. Hess et al., *Pseudo-Complex General Relativity*,
FIAS Interdisciplinary Science Series, DOI 10.1007/978-3-319-25061-8_1

One intuitive way to introduce a new number system involves another well-known mathematical concept: vector spaces over the real numbers. As an example, consider the two-dimensional vector space $\mathbb{R}^2$. Addition and subtraction of vectors, as well as multiplication of vectors with real numbers are well-defined and every physicist is familiar with these concepts, which are often performed using vector components with respect to some basis. What we additionally have to define is a *multiplication of vectors*. Let us use a basis $\{e_1, e_2\}$ of $\mathbb{R}^2$, such that we can write every vector X as $X_1 e_1 + X_2 e_2$. A straightforward way to define a multiplication of two vectors is to write

$$\begin{aligned} X \cdot Y &= (X_1 e_1 + X_2 e_2) \cdot (Y_1 e_1 + Y_2 e_2) \\ &= X_1 Y_1 (e_1 \cdot e_1) + X_1 Y_2 (e_1 \cdot e_2) + X_2 Y_1 (e_2 \cdot e_1) + X_2 Y_2 (e_2 \cdot e_2). \end{aligned} \quad (1.2)$$

Since X_1, X_2, Y_1, Y_2 are *real* numbers, we perform the multiplications $X_1 Y_1$, etc. in the usual way. It follows that we can define multiplication of two arbitrary vectors entirely by providing multiplication rules for the *basis vectors*, for instance choosing

$$\begin{aligned} e_1 \cdot e_1 &:= e_1, \\ e_1 \cdot e_2 &:= e_2 \cdot e_1 := e_2, \\ e_2 \cdot e_2 &:= -e_1. \end{aligned} \quad (1.3)$$

These rules become much more familiar once we write "1" instead of e_1, and "i" instead of e_2. As the reader certainly already noticed, we just have reproduced the complex numbers $\mathbb{C}$. The important point is, that in (1.3) we could also have chosen different multiplication rules, thus obtaining another algebra. Once we have fixed these rules, we have to study the properties of the resulting algebra, looking for a unity element of multiplication, inverse elements, and all that.

Instead of using complex numbers, let us now introduce another algebra, called *pseudo-complex* numbers. Following our previous argument, all we have to do is to provide the multiplication rules for a given basis:

$$\begin{aligned} e_1 \cdot e_1 &:= e_1, \\ e_1 \cdot e_2 &:= e_2 \cdot e_1 := e_2, \\ e_2 \cdot e_2 &:= e_1. \end{aligned} \quad (1.4)$$

We facilitate the notation by writing $e_1 = 1$, and $e_2 = I$, thus also emphasizing the difference to the complex numbers:

$$1 \cdot 1 = 1, \ 1 \cdot I = I \cdot 1 = I, \ I \cdot I = 1. \quad (1.5)$$

These numbers together with the respective multiplication rules are called the pseudo-complex numbers $\mathbb{P}$. For convenience we often omit writing the basis element "1" and use the notation $X_R + I X_I$ for a pseudo-complex number X. In Exercise 1.1 we explicitly show how pseudo-complex numbers are multiplied and divided.

Instead of the set $\{1, I\}$ one can chose an arbitrary linear combination of these elements as the basis for the pseudo-complex numbers. We introduce the basis $\{\sigma_+, \sigma_-\}$, defined by

$$\sigma_+ = \frac{1}{2}(1 + I), \sigma_- = \frac{1}{2}(1 - I),$$

(1.6)

with the inverse relations

$$1 = \sigma_+ + \sigma_-, I = \sigma_+ - \sigma_-.$$

(1.7)

We can write a pseudo-complex number X either as

$$X = X_R + I X_I$$

(1.8)

or

$$X = X_+ \sigma_+ + X_- \sigma_-.$$

(1.9)

Defining the pseudo-complex conjugate of I as $-I$, the pseudo-complex conjugate X^* of X is given by

$$X^* = X_R - I X_I = X_+ \sigma_- + X_- \sigma_+,$$

(1.10)

where we have used $\sigma_\pm^* = \sigma_\mp$. This allows us to define the norm of a variable as

$$|X| = X X^* = \left(X_R^2 - X_I^2\right) = X_+ X_-.$$

(1.11)

It can be shown, that in the $\sigma_\pm$-basis all mathematical operations are performed in the $\pm$-part separately, respectively (see exercises). That is, for two pseudo-complex numbers X and Y we have

$$\begin{aligned} X + Y &= (X_+ \sigma_+ + X_- \sigma_-) + (Y_+ \sigma_+ + Y_- \sigma_-) \\ &= (X_+ + Y_+) \sigma_+ + (X_- + Y_-) \sigma_-, \\ X \cdot Y &= (X_+ \sigma_+ + X_- \sigma_-) \cdot (Y_+ \sigma_+ + Y_- \sigma_-) \\ &= (X_+ Y_+) \sigma_+ + (X_- Y_-) \sigma_-. \end{aligned}$$

(1.12)

(1.13)

This *product structure* is the main feature of the algebra of pseudo-complex numbers. Later we will see that it allows to formulate two copies of General Relativity, which then have to be connected by a new principle to yield a new, modified theory.

We now introduce the concept of *zero divisors* as a "generalized zero" of pseudo-complex numbers. First consider the equation

$$a \cdot b = 0, \qquad (1.14)$$

where a and b are *real* numbers. We know that for real numbers the only solution of this equation is either $a = 0$ or $b = 0$, or zero both. But we have learned that for *pseudo-complex* numbers it holds

$$(X_+\sigma_+) \cdot (Y_-\sigma_-) = X_+Y_- \, (\sigma_- \cdot \sigma_+) = X_+Y_- \cdot 0 = 0. \qquad (1.15)$$

That is, in the algebra of pseudo-complex numbers we have nonzero solutions of the equation

$$X \cdot Y = 0, \qquad (1.16)$$

which belong to the set of so-called *zero divisors*, given by pseudo-complex numbers which have only a component in the σ_+-sector or σ_--sector, respectively:

$$X = X_+\sigma_+ \ \text{or} \ X = X_-\sigma_-. \qquad (1.17)$$

Since these numbers solve an equation, which for the real numbers has only the zero element as a solution, zero divisors can be interpreted as a kind of "generalized zeros". It can be shown, that the zero divisors are just given by the pseudo-complex numbers, which do not have an inverse (see Exercise 1.2). That is, if X is a zero divisor, there is no pseudo-complex number X^{-1} such that $X \cdot X^{-1} = 1$. Since division is defined as multiplication with an inverse element, we obtain the important statement that in

Fig. 1.1 Illustration of the pseudo-complex plane for the variable $X = X_R + I X_I$ $= X_+\sigma_+ + X_-\sigma_-$. The *horizontal* and *vertical* line correspond to the pseudo-real and pseudo-imaginary axes, respectively. The *diagonal lines* represent the zero divisor branch

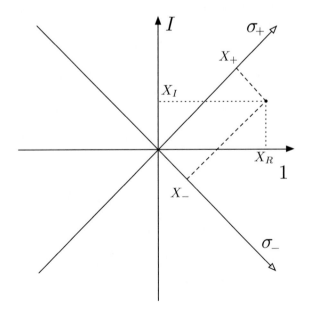

the algebra of pseudo-complex numbers we cannot divide by zero divisors, similar to the real numbers, where we cannot divide by zero. In the following we denote the set of zero divisors by $\mathbb{P}$:

$$\mathbb{P} = \{X \in \mathbb{P} | X = X_+\sigma_+ \text{ or } X = X_-\sigma_-\} = \{X \in \mathbb{P} | X_R^2 = X_I^2\}. \qquad (1.18)$$

In Fig. 1.1 we illustrate the pseudo-complex numbers $\mathbb{P}$ as a two-dimensional plane, with the axes either given by the $\{1, I\}$ or the $\{\sigma_+, \sigma_-\}$ basis elements. A point X in this plane is associated with coordinates (X_R, X_I) or (X_+, X_-), respectively.

Exercise 1.1 (Multiplication and division of pseudo-complex numbers)

Problem.

(a) Calculate explicitly the product XY of two pseudo-complex numbers $X = X_R + IX_I$ and $Y = Y_R + IY_I$.
(b) What is the unity element of multiplication for pseudo-complex numbers?
(c) What is the inverse X^{-1} of a pseudo-complex number X? Does every pseudo-complex number has an inverse?

Solution.

(a)

$$
\begin{aligned}
X \cdot Y &= (X_R 1 + X_I I) \cdot (Y_R 1 + Y_I I) \\
&= X_R Y_R (1 \cdot 1) + X_R Y_I (1 \cdot I) + X_I Y_R (I \cdot 1) + X_I Y_I (I \cdot I) \\
&= (X_R Y_R)1 + (X_R Y_I)I + (X_I Y_R)I + (X_I Y_I)1 \\
&= (X_R Y_R + X_I Y_I) + (X_R Y_I + X_I Y_R) I.
\end{aligned}
$$

Note that the multiplication of two pseudo-complex number is commutative.

(b) The unity element E is determined by the equation $E \cdot Y = Y$ for arbitrary Y. According to the previous calculation we obtain

$$
\begin{aligned}
E \cdot Y &= (E_R Y_R + E_I Y_I) + (E_R Y_I + E_I Y_R) I \\
&\overset{!}{=} Y_R + Y_I I. \qquad (1.19)
\end{aligned}
$$

It follows $E_R = 1$ and $E_I = 0$, and thus the unity element is the basis element, $E = 1$. Note that the unity element is unique in the way that it is the *only* pseudo-complex number such that for every X it holds $E \cdot X = X$.

(c) The inverse X^{-1} of a pseudo-complex number X is determined by the condition $X \cdot X^{-1} = 1$. For convenience denote the inverse of X by Y and consider the calculation in part (a) of this exercise. The condition then reads

$$X_R Y_R + X_I Y_I = 1, \quad X_R Y_I + X_I Y_R = 0. \tag{1.20}$$

Solving the second equation for Y_I, and inserting into the first equation yields

$$X_R Y_R + X_I \left(-\frac{X_I Y_R}{X_R} \right) = 1,$$
$$\Rightarrow X_R^2 Y_R - X_I^2 Y_R = X_R,$$
$$\Rightarrow Y_R = \frac{X_R}{X_R^2 - X_I^2}. \tag{1.21}$$

The second equation can be rewritten as

$$Y_I = -\frac{X_I}{X_R} Y_R, \tag{1.22}$$

and we obtain

$$Y_I = -\frac{X_I}{X_R^2 - X_I^2}. \tag{1.23}$$

The inverse of a pseudo-complex number $X_R + I X_I$ thus is given by

$$X^{-1} = \frac{X_R}{X_R^2 - X_I^2} - I \frac{X_I}{X_R^2 - X_I^2}. \tag{1.24}$$

We immediately see that for all numbers with $X_R^2 = X_I^2$ the inverse X^{-1} is not defined and hence such numbers do not have a multiplicative inverse, which answers (c).

Exercise 1.2 (The $\{\sigma_+, \sigma_-\}$ basis)

Problem.

(a) Derive the relations between the components X_R, X_I and X_+, X_- of a pseudo-complex number X in the $\{1, I\}$ basis and the $\{\sigma_+, \sigma_-\}$ basis, respectively.

(b) The algebra of pseudo-complex numbers is determined by the multiplication rules in (1.5). How do these multiplication rules read in terms of the $\{\sigma_+, \sigma_-\}$ basis?

(c) Calculate explicitly the product of two pseudo-complex numbers in terms of the $\{\sigma_+, \sigma_-\}$ basis. Determine the inverse element of a pseudo-complex number X. Which numbers do not have a multiplicative inverse?

Solution.

(a) Using the definition of the basis elements $\sigma_\pm$ we write

$$
\begin{aligned}
X &= X_+\sigma_+ + X_-\sigma_- \\
&= X_+\frac{1}{2}(1+I) + X_-\frac{1}{2}(1-I) \\
&= \frac{1}{2}(X_+ + X_-) + I\frac{1}{2}(X_+ - X_-).
\end{aligned}
\tag{1.25}
$$

This yields the relations

$$
X_R = \frac{1}{2}(X_+ + X_-), \quad X_I = \frac{1}{2}(X_+ - X_-),
\tag{1.26}
$$

and

$$
X_+ = X_R + X_I, \quad X_- = X_R - X_I.
\tag{1.27}
$$

(b) Using the definition of the basis elements and the multiplication rules for pseudo-complex numbers we obtain

$$
\begin{aligned}
\sigma_+ \cdot \sigma_+ &= \frac{1}{2}(1+I) \cdot \frac{1}{2}(1+I) \\
&= \frac{1}{4}(1 \cdot 1 + 1 \cdot I + I \cdot 1 + I \cdot I) \\
&= \frac{1}{4}(1 + I + I + 1) \\
&= \frac{1}{2}(1+I) = \sigma_+.
\end{aligned}
\tag{1.28}
$$

In the same way we obtain

$$
\begin{aligned}
\sigma_- \cdot \sigma_- &= \frac{1}{2}(1-I) \cdot \frac{1}{2}(1-I) \\
&= \frac{1}{4}(1 - I - I + 1) \\
&= \frac{1}{2}(1-I) = \sigma_-
\end{aligned}
\tag{1.29}
$$

and

$$\sigma_+ \cdot \sigma_- = \frac{1}{2}(1+I) \cdot \frac{1}{2}(1-I)$$
$$= \frac{1}{4}(1+I-I-1)$$
$$= 0. \tag{1.30}$$

(c) Consider two pseudo-complex number X and Y. From the calculations in part (b) of this exercise it follows

$$\begin{aligned} X \cdot Y &= (X_+\sigma_+ + X_-\sigma_-) \cdot (Y_+\sigma_+ + Y_-\sigma_-) \\ &= X_+Y_+(\sigma_+ \cdot \sigma_+) + X_+Y_-(\sigma_+ \cdot \sigma_-) \\ &\quad + X_-Y_+(\sigma_- \cdot \sigma_+) + X_-Y_-(\sigma_- \cdot \sigma_-) \\ &= X_+Y_+\sigma_+ + X_+Y_- \cdot 0 + X_-Y_+ \cdot 0 + X_-Y_-\sigma_- \\ &= X_+Y_+\sigma_+ + X_-Y_-\sigma_-. \end{aligned} \tag{1.31}$$

We observe that in the $\{\sigma_+, \sigma_-\}$ basis multiplication takes place in the $\sigma_\pm$-part separately. Since also addition is performed in both parts separately, in this basis a *product structure* of the pseudo-complex numbers becomes apparent, with the $\sigma_\pm$-parts as two separated sectors. This product structure allows us to determine the inverse immediately. Since the unity element is given by

$$1 = \sigma_+ + \sigma_-, \tag{1.32}$$

the inverse of X has the form

$$X^{-1} = \frac{1}{X_+}\sigma_+ + \frac{1}{X_-}\sigma_-, \tag{1.33}$$

which can be easily checked by calculating

$$\begin{aligned} X \cdot X^{-1} &= (X_+\sigma_+ + X_-\sigma_-) \cdot \left(\frac{1}{X_+}\sigma_+ + \frac{1}{X_-}\sigma_- \right) \\ &= \frac{X_+}{X_+}\sigma_+ + \frac{X_-}{X_-}\sigma_- \\ &= \sigma_+ + \sigma_- = 1. \end{aligned} \tag{1.34}$$

We also observe that for all pseudo-complex numbers with $X_+ = 0$ or $X_- = 0$ the inverse X^{-1} is not defined, and consequently these numbers do not have a multiplicative inverse. Note that this condition is equivalent to the condition $X_R^2 = X_I^2$ obtained in Exercise 1.1.

1.2 Calculus

After introducing pseudo-complex numbers and their basic properties, we now define pseudo-complex functions $f : \mathbb{P} \to \mathbb{P}$ and their derivatives. For convenience in the following we will often abbreviate "pseudo-complex" by "pc", thus speaking of pc-numbers, pc-functions, etc.

A pc-function $f(X)$ maps a pc-number X to another pc-number $f(X)$. Using the $\{1, I\}$ or the $\{\sigma_+, \sigma_-\}$ basis, we represent f by *two real-valued functions with two real arguments, respectively*:

$$
\begin{aligned}
f(X) &= f(X_R + I X_I) \\
&= f_R(X_R, X_I) + I f_I(X_R, X_I) \\
&= f(X_+ \sigma_+ + X_- \sigma_-) \\
&= f_+(X_+, X_-)\sigma_+ + f_-(X_+, X_-)\sigma_- .
\end{aligned}
\tag{1.35}
$$

In the following we will usually write $f_{R,I}(X)$ and $f_\pm(X)$ and keep in mind that these functions depend on the components of X with respect to the basis $\{1, I\}$ and $\{\sigma_+, \sigma_-\}$, respectively.

In many cases real functions are defined as power series. In this case we can adopt the definition of the corresponding pseudo-complex function by replacing the real variable x with a pseudo-complex variable. The separation of the then pseudo-complex function into its $\sigma_\pm$-part assures the convergence of the power series. As an example consider the exponential function $\exp[X]$ defined by the series expansion

$$
\exp[X] = \sum_{n=0}^{\infty} \frac{X^n}{n!} .
\tag{1.36}
$$

Using the basis $\{\sigma_\pm\}$ we can write this as

$$
\begin{aligned}
\exp[X_+ \sigma_+ + X_- \sigma_-] &= \sum_{n=0}^{\infty} \frac{(X_+ \sigma_+ + X_- \sigma_-)^n}{n!} \\
&= \sum_{n=0}^{\infty} \frac{(X_+)^n}{n!}\sigma_+ + \sum_{n=0}^{\infty} \frac{(X_-)^n}{n!}\sigma_- \\
&= \exp[X_+]\sigma_+ + \exp[X_-]\sigma_- ,
\end{aligned}
\tag{1.37}
$$

where we have used the multiplication rule (1.13), and $\exp[X_\pm]$ are the well-known real-valued exponential functions. As is shown in Exercise 1.3, in terms of the $\{1, I\}$ basis the exponential function reads

$$
\exp[X] = \exp[X_R] \left(\cosh[X_I] + I \sinh[X_I]\right),
\tag{1.38}
$$

where $\cosh[X_I]$ and $\sinh[X_I]$ are again real-valued functions.

We now give a definition for differentiability of a pc-function. For simplicity, we always assume that the real functions $f_{R,I}$ and $f_\pm$ are smooth functions with respect to each of its arguments. Then we define the pseudo-complex derivative $f'(X_0)$ at X_0 by

$$f'(X_0) = \left.\frac{Df}{DX}\right|_{X=X_0} = \lim_{X \to X_0} \frac{f(X) - f(X_0)}{X - X_0}.$$
$$\text{with } (X - X_0) \notin \mathbb{P} \tag{1.39}$$

This is analogous to the definition of the derivative of a real function, apart from the additional requirement $(X - X_0) \notin \mathbb{P}$. Recall that $\mathbb{P}$ denotes the set of zero divisors, that is, all pseudo-complex numbers which do not have a multiplicative inverse. So if $X - X_0$ would be a zero divisor, the number $(X - X_0)^{-1}$ would not exist, and we could not perform the division in (1.39). For this reason we can take the limit $X \to X_0$ along an arbitrary path in the pseudo-complex plane $\mathbb{P}$, but have to make sure that we do not cross the set of zero divisors $\mathbb{P}$. Since we want to have the *unique* derivative $f'(X)$ independent from the path we follow in the limit $X \to X_0$, we also have to apply an additional condition on the partial derivatives of the functions $f_{R,I}$ and $f_\pm$. This is completely analogous to the *complex* case, where in this context one speaks of *holomorphic functions* and demands the fulfillment of the Cauchy-Riemann equations. It can be shown [1, 2], that for the pseudo-complex case the argument leads to the following pseudo-complex Cauchy-Riemann equations:

$$\frac{\partial f_R(X)}{\partial X_R} = \frac{\partial f_I(X)}{\partial X_I}, \quad \frac{\partial f_R(X)}{\partial X_I} = \frac{\partial f_I(X)}{\partial X_R}, \tag{1.40}$$

If these equations are fulfilled, the pseudo-complex function f is pseudo-complex differentiable with a unique pc-derivative $f'(X)$ and is called *pseudo-holomorphic*. We can write this derivative in terms of its components (see [1, 2] and Chap. 7 for details):

$$f'(X) = \frac{\partial f_R}{\partial X_R} + I\frac{\partial f_I}{\partial X_R} = \frac{\partial f_I}{\partial X_I} + I\frac{\partial f_R}{\partial X_I}. \tag{1.41}$$

In terms of the $\{\sigma_+, \sigma_-\}$ basis the pseudo-complex Cauchy-Riemann equations have an even simpler form (see Exercise 1.4):

$$\frac{\partial f_+(X)}{\partial X_-} = 0, \quad \frac{\partial f_-(X)}{\partial X_+} = 0. \tag{1.42}$$

It follows the important statement that a pseudo-holomorphic function f can be written as

$$f(X) = f_+(X_+)\sigma_+ + f_-(X_-)\sigma_-, \tag{1.43}$$

with the derivative

$$f'(X) = \frac{\partial f_+}{\partial X_+}\sigma_+ + \frac{\partial f_-}{\partial X_-}\sigma_-. \tag{1.44}$$

That is, for a pseudo-holomorphic function the product structure of pseudo-complex numbers is carried over to its functional properties and its derivative.

After introducing the derivative of a pc-function, as the final piece of this introductory section we now define pseudo-complex integration. Similar to integration in $\mathbb{R}^n$ or $\mathbb{C}$, for the integration of pseudo-complex functions f we have to provide a curve $\gamma : (a,b) \to \mathbb{P}$ to determine the path along we integrate in the pseudo-complex plane. We write

$$\gamma(t) = \gamma_R(t) + I\gamma_I(t), \tag{1.45}$$

where $t \in (a,b)$ is the real-valued parameter. We define pseudo-complex integration by

$$\int_\gamma f\,DX \tag{1.46}$$

with the pseudo-complex differential

$$DX = dX_R + I\,dX_I = dX_+\sigma_+ + dX_-\sigma_-. \tag{1.47}$$

Using the now already familiar rules of pseudo-complex multiplication we get

$$\begin{aligned}
\int_\gamma f\,DX &= \int_\gamma (f_R + f_I I)\,(dX_R + I\,dX_I) \\
&= \int_\gamma (f_R dX_R + f_I dX_I) + I\int_\gamma (f_I dX_R + f_R dX_I) \\
&= \int_\gamma (f_+\sigma_+ + f_-\sigma_-)\,(dX_+\sigma_+ + dX_-\sigma_-) \\
&= \sigma_+\int_\gamma f_+ dX_+ + \sigma_-\int_\gamma f_- dX_-.
\end{aligned} \tag{1.48}$$

Exercise 1.3 (The pseudo-complex exponential function)

Problem. Write the pc-exponential function $\exp[X]$ in terms of the $\{1, I\}$ basis.

Solution. We have shown that

$$\exp[X] = \exp[X_+]\sigma_+ + \exp[X_-]\sigma_-. \tag{1.49}$$

With the relations

$$\sigma_\pm = \frac{1}{2}\,(1 \pm I)\,, \quad X_\pm = X_R \pm X_I \tag{1.50}$$

this reads

$$\exp[X] = \exp[X_R + X_I]\frac{1}{2}\,(1 + I) + \exp[X_R - X_I]\frac{1}{2}\,(1 - I)\,. \tag{1.51}$$

Using $\exp[X_R \pm X_I] = \exp[X_R]\exp[\pm X_I]$ and some rearranging we obtain

$$\exp[X] = \exp[X_R]\Big(\frac{1}{2}\,(\exp[X_I] + \exp[-X_I]) + I\frac{1}{2}\,(\exp[X_I] - \exp[-X_I])\Big)\,. \tag{1.52}$$

Since $\exp[X_I]$ here is a *real* function, we can use the definitions

$$\cosh[x] = \frac{1}{2}\,(\exp[x] + \exp[-x])\,, \quad \sinh[x] = \frac{1}{2}\,(\exp[x] - \exp[-x])\,, \tag{1.53}$$

and finally obtain

$$\exp[X] = \exp[X_R]\Big(\cosh[X_I] + I\,\sinh[X_I]\Big)\,. \tag{1.54}$$

Exercise 1.4 (The pseudo-complex Cauchy-Riemann equations)

Problem. Which relations for the partial derivatives of $f_\pm$ with respect to $X_\pm$ follow from the pseudo-complex Cauchy Riemann equations (1.40)?

Solution. The pseudo-complex Cauchy Riemann relations read

$$\frac{\partial f_R(X_R, X_I)}{\partial X_R} = \frac{\partial f_I(X_R, X_I)}{\partial X_I}\,, \quad \frac{\partial f_R(X_R, X_I)}{\partial X_I} = \frac{\partial f_I(X_R, X_I)}{\partial X_R}\,, \tag{1.55}$$

where we have explicitly written the dependency on X_R and X_I. Recall the following relations:

$$f_\pm = f_R \pm f_I\,, \quad X_R = \frac{1}{2}\,(X_+ + X_-)\,, \quad X_I = \frac{1}{2}\,(X_+ - X_-)\,. \tag{1.56}$$

Using these relations, we can write

$$
\begin{aligned}
\frac{\partial f_+(X_+, X_-)}{\partial X_+} &= \frac{\partial}{\partial X_+} f_R\Big(X_R(X_+, X_-), X_I(X_+, X_-)\Big) \\
&\quad + \frac{\partial}{\partial X_+} f_I\Big(X_R(X_+, X_-), X_I(X_+, X_-)\Big) \\
&= \frac{\partial f_R(X_R, X_I)}{\partial X_R}\frac{\partial X_R}{\partial X_+} + \frac{\partial f_R(X_R, X_I)}{\partial X_I}\frac{\partial X_I}{\partial X_+} \\
&\quad + \frac{\partial f_I(X_R, X_I)}{\partial X_R}\frac{\partial X_R}{\partial X_+} + \frac{\partial f_I(X_R, X_I)}{\partial X_I}\frac{\partial X_I}{\partial X_+} \\
&= \frac{1}{2}\frac{\partial f_R(X_R, X_I)}{\partial X_R} + \frac{1}{2}\frac{\partial f_R(X_R, X_I)}{\partial X_I} \\
&\quad + \frac{1}{2}\frac{\partial f_I(X_R, X_I)}{\partial X_R} + \frac{1}{2}\frac{\partial f_I(X_R, X_I)}{\partial X_I} \\
&= \frac{\partial f_R(X_R, X_I)}{\partial X_R} + \frac{\partial f_R(X_R, X_I)}{\partial X_I}.
\end{aligned} \tag{1.57}
$$

In the last line we have used the pseudo-complex Cauchy-Riemann equations, and thus could also have written the same expression with R and I interchanged. Performing an analogous calculation we get

$$
\begin{aligned}
\frac{\partial f_+(X_+, X_-)}{\partial X_-} &= \frac{\partial}{\partial X_-} f_R\Big(X_R(X_+, X_-), X_I(X_+, X_-)\Big) \\
&\quad + \frac{\partial}{\partial X_-} f_I\Big(X_R(X_+, X_-), X_I(X_+, X_-)\Big) \\
&= \frac{\partial f_R(X_R, X_I)}{\partial X_R}\frac{\partial X_R}{\partial X_-} + \frac{\partial f_R(X_R, X_I)}{\partial X_I}\frac{\partial X_I}{\partial X_-} \\
&\quad + \frac{\partial f_I(X_R, X_I)}{\partial X_R}\frac{\partial X_R}{\partial X_-} + \frac{\partial f_I(X_R, X_I)}{\partial X_I}\frac{\partial X_I}{\partial X_-} \\
&= \frac{1}{2}\frac{\partial f_R(X_R, X_I)}{\partial X_R} - \frac{1}{2}\frac{\partial f_R(X_R, X_I)}{\partial X_I} \\
&\quad + \frac{1}{2}\frac{\partial f_I(X_R, X_I)}{\partial X_R} - \frac{1}{2}\frac{\partial f_I(X_R, X_I)}{\partial X_I} \\
&= 0.
\end{aligned} \tag{1.58}
$$

In the same way we also get

$$
\frac{\partial f_-(X_+, X_-)}{\partial X_-} = \frac{\partial f_R(X_R, X_I)}{\partial X_R} - \frac{\partial f_R(X_R, X_I)}{\partial X_I} \tag{1.59}
$$

and

$$\frac{\partial f_-(X_+, X_-)}{\partial X_+} = 0. \tag{1.60}$$

It follows that for a pseudo-holomorphic function f we can write

$$f = f_+(X_+)\sigma_+ + f_-(X_-)\sigma_-. \tag{1.61}$$

References

1. F. Antonuccio, *Semi-Complex Analysis and Mathematical Physics* (1993), http://arxiv.org/abs/gr-qc/9311032v2
2. P.M. Gadea, J. Grifone, J. Muñoz, Masqué. Manifolds modelled over free modules over the double numbers. Acta Math. Hungar **100**, 187 (2003)

Chapter 2
Pseudo-complex General Relativity

The theory of General Relativity (GR) has passed numerous high precision tests, from the measured perihelion shift of Mercury, indirect hints to gravitational waves up to frame dragging effects (Frame dragging effects will be discussed in Chap. 5. The effect itself refers to particles which orbit a central mass counter-clockwise in a weak gravitational field, but which are *dragged* to rotate clock-wise in the extreme situation of a very strong gravitational field.). GR also implies the existence of so-called black holes, which even trap light. Such massive objects are claimed to exist in the center of almost any galaxy. Physics nearby such objects is investigated and GR is up to now the only theory which is claimed to describe the phenomena around these dense objects. The existence of black holes has become commonly accepted despite the fact that the existence of event horizons cannot be proved from observational data [1].

Having a successful theory one may ask the question: Why to search for extensions/corrections? The answer is simple: From a theoretical point of view, GR has still problems. For example, the black hole consists of a singularity at the center of a very compact mass distribution, which is hidden by an event horizon. A sufficiently far observer can not see inside (though an astronaut falling into a black hole can pass the event horizon until he is shredded by huge tidal forces). For a nearby observer (at a save distance) the region marked by the event horizon is excluded from observation: it cannot be accessed by him. This is from a philosophical point unsatisfactory. No part of the (nearby) space should be excluded from physical studies!

Furthermore, GR has not been tested in extremely strong gravitational fields, as they appear near the Schwarzschild radius, the radial distance of the event horizon from the singularity. All effects observed near massive objects correspond to distances of several Schwarzschild radii, where the gravitational field is still strong but not as intense as at the distance of one Schwarzschild radius. Therefore, there is no valid argument that GR should continue to be correct when the Schwarzschild radius is approached [1]. Also, GR may be contained in a more general theory, which should allow deviations.

© Springer International Publishing Switzerland 2016
P.O. Hess et al., *Pseudo-Complex General Relativity*,
FIAS Interdisciplinary Science Series, DOI 10.1007/978-3-319-25061-8_2

There are other reasons why in the past several attempts have been made to extend GR, for example to unify GR with Electrodynamics [2, 3], eliminate inconsistencies with Quantum Mechanics [4, 5] or to introduce effects of a minimal length [6]. All are related to algebraical extensions of the theory, using instead of real coordinates complex ones or more general constructions. Our motivation is based on the leading principle that a theory should have *no singularity*. *As long as in theoretical models singularities appear, these models are incomplete.* This holds not only for General Relativity but for any theory, i.e., also for QED and QCD! Although it is no trivial task to define a singularity in General Relativity in a precise and coordinate-invariant way [7, 8], there are at least two well-known singularities related to non-hypothetical, but rather concrete physical systems. This is on the one hand the singularity at the beginning of the universe, and on the other hand the (hidden) singularity at the center of a so-called black hole. In the latter case, an event horizon, which in most cases can be approximated by a surface from which no signal can escape, is supposed to hide the singularity at the center. We seek for a form of GR for which the singularity of GR, i.e. in particular for the Schwarzschild singularity, at the center for large, highly concentrated masses can be avoided and also whether the event horizon can be eliminated. If this is the case, then the massive objects, called up to now black holes, would disappear and instead rather dark areas (or *dark stars*) would occur. Though they mimic a black hole, looking at them from far away, in reality one would be able to study their structure.

In the following sections, we will shortly describe attempts of modifying GR and finally propose ours, called the pseudo-complex General Relativity (pc-GR).

2.1 The Attempt by A. Einstein

The motivation of A. Einstein was to search for a *unified field theory*, in particular of GR and Electrodynamics [2, 3]. He mentions as an example of such a unification the theory of Special Relativity, especially the relativistic transformation of electric to magnetic fields and vice versa. In fact, electric and magnetic fields can not be treated independently but are resumed into the skew symmetric field tensor $(F_{\mu\nu})$, whose matrix elements are related to either electric or magnetic field components. Applying a Lorentz transformation converts them, one into the other. A similar phenomenon he expected for a unified field theory of GR and Electrodynamics. This is an example, that there may exist different motivations to extend GR, not only related to get rid of the event horizon.

The analog to the electromagnetic tensor $F_{\mu\nu} = -F_{\nu\mu}$ is the metric tensor $g_{\mu\nu} = g_{\nu\mu}$ in General Relativity. For the unified field theory, Einstein proposed [2, 3] to extend the metric tensor to a complex field

$$G_{\mu\nu} = g_{\mu\nu} + i F_{\mu\nu} \tag{2.1}$$

which satisfies the *hermitian* condition

$$G^*_{\mu\nu} = G_{\nu\mu}. \tag{2.2}$$

Using (2.1) and (2.2) leads to the conclusion that the real part is symmetric and the imaginary part has to be anti-symmetric. The real part represents the metric in standard GR while the imaginary part is attributed to the electromagnetic field tensor. In such a way, an apparent unification is obtained.

In order to define the parallel displacement of complex vectors, Einstein had to introduce complex connection coefficients $\Gamma^\lambda_{\mu\nu}$, which are hermitian with respect to the lower indices. This shows some similarity with a theory based on real, but asymmetric connection coefficients (torsion). We will not go into the detail here, but rather refer to our chapter on the geometric differential formulation (Chap. 7), which shows similarities to Einstein's extended theory.

It should be noted, that Einstein did not introduce complex coordinates, but only complex fields defined for real variables. The same approach was taken in a mathematically more elaborate way by Kunstatter and Yates [9]. Nevertheless, they made the remark that in principle one could also consider complex coordinates

$$X^\mu = X^\mu_R + i X^\mu_I, \tag{2.3}$$

which would lead to additional degrees of freedom. For instance, one would have to define not only the parallel displacement of complex vectors under real, but also under complex coordinate displacement, thus doubling the degrees of freedom for the connection coefficients compared to the extended theory as proposed by Einstein. We will describe such a theory in the later section on Hermitian Gravity.

2.2 Caianiello's Observation

E.R. Caianiello investigated the properties of the length element

$$d\omega^2 = g_{\mu\nu} \left[dx^\mu dx^\nu + dy^\mu dy^\nu \right]. \tag{2.4}$$

where he identified the components y^μ with the four-velocity, The four-velocity u^μ is defined here as $\frac{dx^\mu}{ds} = \frac{dx^\mu}{cd\tau}$, with τ denoting the eigentime. This definition is due to the common notation in literature.

$$y^\mu = l u^\mu \rightarrow dy^\mu = l du^\mu. \tag{2.5}$$

The parameter l is introduced in order to maintain the units of length. That is, differently to the approach by Einstein, he did not consider some extended algebraic (complex) structure on real coordinates, but rather extended the coordinates to eight dimensions, with the new coordinates given by the respective four-velocities. He then considered the real space as embedded into this generalized space, with the procedure leading to an additional contribution based on accelerations.

In terms of the coordinates and the 4-velocity, the length element is

$$d\omega^2 = g_{\mu\nu}\left[dx^\mu dx^\nu + l^2 du^\mu du^\nu\right].\tag{2.6}$$

In standard GR, the length element along a trajectory is given by $ds^2 = g_{\mu\nu}dx^\mu dx^\nu$. For simplicity, we consider for a moment a flat space, with $g_{\mu\nu} = \eta_{\mu\nu}$ and $\eta_{\mu\nu} = \text{diag}(-1, +1, +1, +1)$. Using that $ds^2 = -c^2 d\tau^2 = g_{\mu\nu}dx^\mu dx^\nu$, where τ is the proper time, one defines the 4-acceleration components as

$$a^\mu = \frac{du^\mu}{(-c)d\tau} = \frac{d^2x^\mu}{ds^2}.\tag{2.7}$$

we have

$$-\eta_{\mu\nu}a^\mu a^\nu = a^0 a^0 - a^i a^i = -a^2.\tag{2.8}$$

This is the negative of the acceleration squared, where the sign is taken from the second term (the spatial components). This identification with the 4-velocity is generalized to a curved space ($g_{\mu\nu} \neq \eta_{\mu\nu}$). Identifying the first term in (2.6) with ds^2, the length element can be rewritten as

$$d\omega^2 = \left[1 - l^2 a^2\right]ds^2 = \left[1 - l^2 a^2\right]g_{\mu\nu}dx^\mu dx^\nu.\tag{2.9}$$

This corresponds to a modified metric

$$G_{\mu\nu} = \left[1 - l^2 a^2\right]g_{\mu\nu}\tag{2.10}$$

due to the acceleration.

Considering that a particle moves only along world lines with a negative length square (positive proper time, otherwise the proper time passed would be imaginary), this implies that the correction factor in (2.9) has to be positive, or

$$a^2 \leq \frac{1}{l^2},\tag{2.11}$$

which corresponds to a maximal acceleration, or equivalently to a minimal length. A beautiful property of this approach is that the minimal length l is a *parameter not affected by Lorentz transformations*. Thus, all continuous symmetries are maintained and there is no need to deform them such that a *physical* minimal length is constant.

An interesting consequence is obtained when $d\omega^2$ is identified with an *extended proper time* $-c^2 d\tilde{\tau}^2$. Let us consider again a flat space ($g_{\mu\nu} = \eta_{\mu\nu}$). With the identification of the proper time we have

$$-c^2 d\tilde{\tau}^2 = \left[1 - l^2 a^2\right]\left(-c^2 dt^2 + dx^2 + dy^2 + dz^2\right)$$

$$= -\left[1 - l^2 a^2\right]\left[1 - \frac{v^2}{c^2}\right] dt^2, \tag{2.12}$$

were v is the velocity of the system moving with respect to an observer, or

$$dt = \frac{d\tilde{\tau}}{\sqrt{1 - (la)^2}\sqrt{1 - \frac{v^2}{c^2}}}. \tag{2.13}$$

This is nothing but the time dilation of two system, moving with respect to each other. There is the usual contribution due to the relative velocity v and in addition a correction due to the acceleration. This implies that the observation depends on the state of motion and thus the strong principle of GR, that gravitational and accelerated systems can not be distinguished by a local observer, is violated. Consequences should be worked out. However, due to the smallness of l (probably of the order of the Planck length of 10^{-33} cm) it is not possible up to now to measure deviations.

Exercise 2.1 (Maximal acceleration in the Schwarzschild metric)

Problem. Consider Caianiello's modified Lagrangian for a spherical symmetric metric, e.g. as in the Schwarzschild case (see also (2.9)), using now the definition that a dot refers to the derivative with respect to ω.

$$L = \frac{d\omega^2}{d\omega^2} = 1 = \Sigma^2 g_{\mu\nu} \dot{x}^\mu \dot{x}^\nu$$

$$= \Sigma^2 \left\{ -e^\nu \dot{t}^2 + e^\lambda \dot{r}^2 + r^2 \left(\dot{\vartheta}^2 + \sin^2 \vartheta \, \dot{\varphi}^2 \right) \right\}, \tag{2.14}$$

with

$$\Sigma^2 = \left(1 + \left(\frac{l}{c}\right)^2 g_{\mu\nu} \ddot{x}^\mu \ddot{x}^\nu\right). \tag{2.15}$$

Restrict to the equatorial plane motion ($\vartheta = \frac{\pi}{2}$, $\dot{\vartheta} = 0$) and use $\lambda = -\nu$.

Show that an equation can be obtained for the determination of Σ^2, which is a function on r and conservative quantities.

The steps, presented here, are due to [10, 11] within the standard GR.

Solution. The partial derivatives of the Lagrangian with respect to the coordinates t, ϕ and their velocities are

$$\frac{\partial L}{\partial \dot{t}} = -2\Sigma^2 e^\nu \dot{t}, \quad \frac{\partial L}{\partial t} = 0,$$

$$\frac{\partial L}{\partial \dot{\varphi}} = 2\Sigma^2 r^2 \dot{\varphi}, \quad \frac{\partial L}{\partial \varphi} = 0. \tag{2.16}$$

This results into the Euler-Lagrange equations

$$\frac{d}{d\omega}\frac{\partial L}{\partial \dot{t}} - \frac{\partial L}{\partial t} = \frac{d}{d\omega}\left(-2\Sigma^2 e^\nu \dot{t}\right) = 0,$$

$$\frac{d}{d\omega}\frac{\partial L}{\partial \dot{\varphi}} - \frac{\partial L}{\partial \varphi} = \frac{d}{d\omega}\left(2\Sigma^2 r^2 \dot{\varphi}\right) = 0. \tag{2.17}$$

These equations give us conserved quantities, defined as

$$E = \Sigma^2 e^\nu \dot{t}, \quad L = -\Sigma^2 r^2 \dot{\varphi}. \tag{2.18}$$

In order to get an expression for $\dot{r}$, it is preferable not to use the corresponding Euler-Lagrange equation, but rather to resolve (2.14) for $\dot{r}^2$ and substitute for the velocities $\dot{t}$ and $\dot{\varphi}$ by (2.18). With the Lagrangian $L = 1$ one gets

$$1 = \Sigma^2 \left[-e^\nu \dot{t}^2 + e^{-\nu} \dot{r}^2 + r^2 \dot{\varphi}^2\right], \tag{2.19}$$

where we have used $\lambda = -\nu$. Substituting $\dot{t}$ and $\dot{\varphi}$ by the conserved quantities we obtain

$$1 = \Sigma^2 \left[-\frac{e^{-\nu} E^2}{\Sigma^4} + e^{-\nu} \dot{r}^2 + \frac{L^2}{\Sigma^4 r^2}\right]. \tag{2.20}$$

Resolving for $\dot{r}^2$ gives

$$\dot{r}^2 = \frac{1}{\Sigma^4}\left\{E^2 + e^\nu \left[\Sigma^2 - \frac{L^2}{r^2}\right]\right\}. \tag{2.21}$$

Next, the accelerations of t and ϕ are determined:

$$\ddot{t} = -Ee^{-\nu}\left\{\frac{1}{\Sigma^4}\frac{\partial \Sigma^2}{\partial r} + \frac{1}{\Sigma^2}\frac{\partial \nu}{\partial r}\right\}\dot{r},$$

$$\ddot{\varphi} = L\left\{\frac{1}{r^2 \Sigma^4}\frac{\partial \Sigma^2}{\partial r} + \frac{2}{r^3 \Sigma^2}\right\}\dot{r}. \tag{2.22}$$

For the acceleration of r we use

$$\frac{d}{d\omega}\dot{r}^2 = 2\dot{r}\ddot{r}$$

$$= \left\{-\frac{2}{\Sigma^6}\frac{\partial \Sigma^2}{\partial r}\left[E^2 + e^\nu\left(\Sigma^2 - \frac{L^2}{r^2}\right)\right]\right.$$

$$\left. +\frac{\partial \nu}{\partial r}\frac{e^\nu}{\Sigma^4}\left[\Sigma^2 - \frac{L^2}{r^2}\right] - \frac{e^\nu}{\Sigma^4}\left[\frac{\partial \Sigma^2}{\partial r} - \frac{2L^2}{r^3}\right]\right\}\dot{r}, \quad (2.23)$$

which yields

$$\ddot{r} = -\left\{\left[\frac{E^2}{\Sigma^6} + e^\nu\left[\frac{1}{2\Sigma^4} - \frac{L^2}{r^2\Sigma^6}\right]\right]\frac{\partial \Sigma^2}{\partial r}\right.$$

$$\left. +\frac{e^\nu}{2\Sigma^4}\left[\frac{\partial \nu}{\partial r}\left(\Sigma^2 - \frac{L^2}{r^2}\right) + \frac{2L^2}{r^3}\right]\right\}. \quad (2.24)$$

The expressions for the accelerations have to be substituted into

$$F(r, \Sigma^2) = g_{\mu\nu}\ddot{x}^\mu \ddot{x}^\nu$$
$$= -e^\nu(\ddot{t})^2 + e^{-\nu}(\ddot{r})^2 + r^2(\ddot{\varphi})^2. \quad (2.25)$$

One then has to solve the equation

$$\Sigma^2 = \left(1 + \left(\frac{l}{c}\right)^2 F(r, \Sigma^2)\right). \quad (2.26)$$

Since this equation can not be solved in closed form, one has to apply an iteration scheme. The first step is to calculate the second derivatives for the case $\Sigma^2_{(0)} = 1$, thus neglecting the effects of a maximal acceleration. Using this first iteration, one then can calculate the new value for Σ^2:

$$\Sigma^2_{(1)} = \left(1 + \left(\frac{l}{c}\right)^2 F(r, \Sigma^2_{(0)})\right). \quad (2.27)$$

This new value $\Sigma^2_{(1)}$ then is used to determine the updated second derivatives, using the functions derived above. It has been shown that already a first iteration leads to interesting results [10, 11].

The length square element (2.6) is the starting point of several other contributions [12–18]. In [16–18] the geometric differential formulation of this modified theory is given. A difficulty arises due to the fact that $u^\mu = \frac{dx^\mu}{ds}$ is a tangential vector on the world line, i.e., it is an element of the so-called *tangent space* (see Chap. 7 on the geometric differential formulation) and, thus, transforms differently than x^μ. Due to that in [16–18] the transformations related to the combined spaces (in the manifold given by (x^μ) and the tangent space given by (u^μ)) acquire a very complicated structure. In [12–15] further properties of the modified theory are worked out, relating it to a Finsler space (which we will not discuss explicitly here).

Interesting to mention is [19], where the invariant properties of the length square element is investigated and the group theory is discussed.

2.3 Concerns Brought Forward by M. Born

The main motivation of M. Born was to unify Quantum Mechanics with GR, a problem which remains unresolved until today. M. Born observed an inconsistency between Quantum Mechanics and General Relativity, which he tried to circumvent in [4, 5]. His idea will be sketched in what follows.

In Quantum Mechanics the position and momentum operators satisfy the commutation relations

$$\left[x^\mu, p_\nu\right] = i\hbar\delta^\mu_\nu, \quad \left[x^\mu, x^\nu\right] = 0, \quad \left[p^\mu, p^\nu\right] = 0, \tag{2.28}$$

where $x^0 = ct$ and $p_0 = \frac{\hbar}{i}\frac{\partial}{\partial(ct)}$. Obviously, the coordinates and momenta have a symmetrical relation: Applying a canonical transformation, interchanging the coordinates with the momenta, i.e.,

$$x^\mu \rightarrow p^\mu, \quad p^\mu \rightarrow -x^\mu, \tag{2.29}$$

the above equations maintain their structure. M. Born denoted it a *reciprocal relation*. (Born notes [4] that "the word" "reciprocity" is chosen because it is already generally used in the lattice theory of crystals where the motion of the particle is described in the p-space with help of the "reciprocal lattice" [sic]).

The situation is different in GR. There the dominant object is the length element square $ds^2 = g_{\mu\nu}dx^\mu dx^\nu$. There is no place for the momenta, which are treated distinctly.

M. Born proposed a modification, namely

$$d\omega^2 = g_{\mu\nu}dx^\mu dx^\nu + \left(\frac{l}{mc}\right)^2 g_{\mu\nu}dp^\mu dp^\nu, \tag{2.30}$$

(actually M. Born uses in general different metrics for the x and p component, but for simplicity it is not done here) with p^μ being the contravariant component of the momentum and m the mass of a particle under consideration. The factor before the metric in the second term serves to maintain the units. Note, that this formulation is glued to a specific particle with mass m. This contradicts the convention that the length element square is a geometric object and should be independent of the particle. A way around is to use 4-velocities, i.e.,

$$d\omega^2 = g_{\mu\nu}dx^\mu dx^\nu + l^2 g_{\mu\nu}du^\mu du^\nu. \tag{2.31}$$

The 4-velocity can be considered as the tangent vector with respect to a given point (x^μ) in the coordinate manifold and it can be arbitrary. Only later on, when the motion of a particle is considered, the components of the 4-velocity acquire specific values. Note also, that the length element square has the same form as in (2.6) and, thus, all consequences will be the same, like the appearance of a maximal acceleration. According to M. Born, due to the smallness of the minimal length l, the second term in the modified length element square contributes only in the *small* world, when momenta are large, while in the *big* world the standard form is recovered, because the second term can then be neglected.

2.4 Hermitian Gravity

Several of the ideas presented above are united into the *Hermitian Gravity* [20, 21]. A further basic ingredient is the algebraic extension to complex coordinates $X^\mu = X_R^\mu + iX_I^\mu$. For the complex conjugate we use the equivalent notations $(X^\mu)^* = \overline{X}^\mu = X_R^\mu - iX_I^\mu$. Sometimes it is convenient to use $(X^\mu, \overline{X}^\mu)$ as the set of independent coordinates instead of (X_R^μ, X_I^μ). We presented in a previous chapter Einstein's idea of a complexified theory of general relativity. He used complex fields for real coordinates, and thus doubled the degrees of freedom for the metric, introducing a real and an imaginary part. These degrees of freedom were then reduced by demanding the symmetry of the real part, and the antisymmetry of the imaginary part. For a more general theory, which also considers complex coordinates, we have even more degrees of freedom. This is due to the introduction of complex coordinate displacements like

$$dX^\mu = dX_R^\mu + idX_I^\mu. \tag{2.32}$$

Using the coordinates $(X^\mu, \overline{X}^\mu)$ the generalized complex line element is then written as

$$d\omega^2 = G_{\mu\nu}dX^\mu dX^\nu + G_{\mu\bar\nu}dX^\mu d\overline{X}^\nu + G_{\bar\mu\nu}d\overline{X}^\mu dX^\nu + G_{\bar\mu\bar\nu}d\overline{X}^\mu d\overline{X}^\nu. \tag{2.33}$$

The bar over the indices μ, ν refer to the index of the components of the pc-conjugate coordinates $\overline{X}^\mu$, $\overline{X}^\nu$. This corresponds to four complex-valued metric tensors, thus eight times the degrees of freedom for the metric tensor in real four-dimensional space-time. Using a more formal language, this approach corresponds to a complex-ification of the tangent and cotangent space of a differential manifold with complex coordinates, and the introduction of a complex metric on this structure. See Chap. 7 for a similar procedure for the corresponding pseudo-complex objects.

The first thing which is demanded is the symmetry of this complex metric, that is

$$G_{\mu\nu} = G_{\nu\mu}, \ G_{\mu\bar{\nu}} = G_{\bar{\nu}\mu}, \ G_{\bar{\mu}\nu} = G_{\nu\bar{\mu}}, \ G_{\bar{\mu}\bar{\nu}} = G_{\bar{\nu}\bar{\mu}}. \tag{2.34}$$

The next step is to assume a real line element, that is

$$d\omega^2 = \left(d\omega^2\right)^*. \tag{2.35}$$

Since this condition has to hold for arbitrary complex coordinate displacements, we obtain the conditions

$$(G_{\mu\nu})^* = G_{\bar{\mu}\bar{\nu}}, \ (G_{\mu\bar{\nu}})^* = G_{\bar{\mu}\nu}. \tag{2.36}$$

There are two different kinds of gravities considered in the literature, depending on the definition of the length element square. In [9, 22] a rather technical definition is given, which can be understood having read our later Chap. 7. Here we use a more pedestrian method, though completely equivalent to the before mentioned theories.

(a) Hermitian metric: For the Hermitian metric one sets

$$G_{\mu\nu} = G_{\bar{\mu}\bar{\nu}} = 0. \tag{2.37}$$

The metric has the structure of an 8×8 matrix, namely

$$(G_{\alpha\beta}) = \begin{pmatrix} 0 & G_{\mu\bar{\nu}} \\ G_{\bar{\mu}\nu} & 0 \end{pmatrix}, \tag{2.38}$$

with $G_{\bar{\mu}\nu} = (G_{\mu\bar{\nu}})^*$. The indices μ and ν run from 1 to 4 and $\bar{\mu}$ and $\bar{\nu}$ also run in the same interval. The α and β run from 1 to 8.

This yields the length element square

$$d\omega^2 = G_{\mu\bar{\nu}} dX^\mu d\overline{X}^\nu + G_{\bar{\mu}\nu} d\overline{X}^\mu dX^\nu \tag{2.39}$$

We now write

$$G_{\mu\bar{\nu}} = \frac{1}{2} \left(g^R_{\mu\nu} + ig^I_{\mu\nu}\right), G_{\bar{\mu}\nu} = (G_{\mu\bar{\nu}})^* = \frac{1}{2} \left(g^R_{\mu\nu} - ig^I_{\mu\nu}\right), \tag{2.40}$$

and using the definition of X^μ and $\overline{X}^\mu$ we obtain

$$
\begin{aligned}
d\omega^2 &= \frac{1}{2}\left(g_{\mu\nu}^R + ig_{\mu\nu}^I\right)\left\{dX_R^\mu dX_R^\nu - idX_R^\mu dX_I^\nu + idX_I^\mu dX_R^\nu + dX_I^\mu dX_I^\nu\right\} \\
&\quad + \frac{1}{2}\left(g_{\mu\nu}^R - ig_{\mu\nu}^I\right)\left\{dX_R^\mu dX_R^\nu + idX_R^\mu dX_I^\nu - idX_R^\mu dX_I^\nu + dX_I^\mu dX_I^\nu\right\} \\
&= g_{\mu\nu}^R\left\{dX_R^\mu dX_R^\nu + dX_I^\mu dX_I^\nu\right\} + g_{\mu\nu}^I\left\{dX_R^\mu dX_I^\nu - dX_I^\mu dX_R^\nu\right\}.
\end{aligned}
\tag{2.41}
$$

Using the symmetries $G_{\bar\nu\mu} = G_{\mu\bar\nu}$ and $G_{\bar\mu\nu} = G_{\nu\bar\mu}$ we can write

$$
g_{\nu\mu}^I = i\left(G_{\bar\nu\mu} - G_{\nu\bar\mu}\right) = i\left(G_{\mu\bar\nu} - G_{\bar\mu\nu}\right) = -g_{\mu\nu}^I.
\tag{2.42}
$$

and observe that $g_{\mu\nu}^I$ is antisymmetric. In the same way one can show that $g_{\mu\nu}^R$ is symmetric. We thus obtain the line element square

$$
d\omega^2 = g_{\mu\nu}^R\left\{dX_R^\mu dX_R^\nu + dX_I^\mu dX_I^\nu\right\} + 2g_{\mu\nu}^I dX_R^\mu dX_I^\nu.
\tag{2.43}
$$

As shown in [23], this theory suffers from ghost or tachyon solutions (see more detailed remarks in Exercise 2.2), a reason why also Einstein's unified theory was not further pursued.

(b) Anti-hermitian metric:

The Anti-hermitian metric is defined by

$$
G_{\mu\bar\nu} = G_{\bar\mu\nu} = 0.
\tag{2.44}
$$

In matrix form this reads

$$
(G_{\alpha\beta}) = \begin{pmatrix} G_{\mu\nu} & 0 \\ 0 & G_{\bar\mu\bar\nu} \end{pmatrix}.
\tag{2.45}
$$

The length element square reads

$$
d\omega^2 = G_{\mu\nu}dX^\mu dX^\nu + G_{\bar\mu\bar\nu}d\overline{X}^\mu d\overline{X}^\nu.
\tag{2.46}
$$

Similar to the case of the hermitian metric we write

$$
G_{\mu\nu} = \frac{1}{2}\left(g_{\mu\nu}^R + ig_{\mu\nu}^I\right), \ \ G_{\bar\mu\bar\nu} = (G_{\mu\nu})^* = \frac{1}{2}\left(g_{\mu\nu}^R - ig_{\mu\nu}^I\right).
\tag{2.47}
$$

This leads to

$$
\begin{aligned}
d\omega^2 &= \frac{1}{2}\left(g_{\mu\nu}^R + ig_{\mu\nu}^I\right)\left\{dX_R^\mu dX_R^\nu + idX_R^\mu dX_I^\nu + idX_I^\mu dX_R^\nu - dX_I^\mu dX_I^\nu\right\} \\
&\quad + \frac{1}{2}\left(g_{\mu\nu}^R - ig_{\mu\nu}^I\right)\left\{dX_R^\mu dX_R^\nu - idX_R^\mu dX_I^\nu - idX_I^\mu dX_R^\nu - dX_I^\mu dX_I^\nu\right\} \\
&= g_{\mu\nu}^R\left\{dX_R^\mu dX_R^\nu - dX_I^\mu dX_I^\nu\right\} - g_{\mu\nu}^I\left\{dX_R^\mu dX_I^\nu + dX_I^\mu dX_R^\nu\right\} \\
&= g_{\mu\nu}^R\left\{dX_R^\mu dX_R^\nu - dX_I^\mu dX_I^\nu\right\} - 2g_{\mu\nu}^I dX_R^\mu dX_I^\nu.
\end{aligned}
\tag{2.48}
$$

In the last step we used that for the anti-hermitian metric both $g_{\mu\nu}^R$ and $g_{\mu\nu}^I$ are symmetric. In [21] it is shown that this theory does not suffer from ghost solutions. However, the form of the length element square is rather strange, since it is, due to the negative sign, not invariant under an orthogonal transformation in the 8-dimensional space but rather under a hyperbolic transformation. This line element does not have a maximal acceleration, as in the former case and as exposed further above.

2.5 The Approach in pc-GR

Philosophically we remember Dirac's procedure of taking the square of

$$
E = \pm\sqrt{p^2c^2 + m_0^2c^4},
\tag{2.49}
$$

with m_0 as the rest mass of a particle, in a different—at that time unconventional way—namely

$$
\mathbf{E} = \gamma^\mu \mathbf{p}_\mu,
\tag{2.50}
$$

where γ^μ are the well known Dirac matrices. This leads to the Dirac equation, yielding spin for the particles (in particular spin $\frac{1}{2}$ for the electrons), the predictions of anti-particles and a model for the vacuum. It shows that new concepts may lead to new physics. Our goal is similar: We want to have *no singularities* (no black holes) in General Relativity with all its consequences. Of course, the new theory should reproduce all experimental data so far obtained but show essential differences in very strong gravitational fields. As we will see, a new *Weltbild* (view of the world) will emerge!

As in [22], pseudo-complex variables (called in the mentioned publication as hyperbolic) are introduced. The algebraic extended variables are given by

$$
X^\mu = x^\mu + Iy^\mu = x^\mu + Ix^{\bar\mu},
\tag{2.51}
$$

where in the second equation an alternative notation for the pseudo-imaginary component is introduced, which will be of use in an 8-dimensional formulation. The index

"μ" refers to the first four components while "$\bar{\mu}$" does to the last four components
(5–8).

The length element square is defined as (use the rules explained in Chap. 1)

$$
\begin{aligned}
d\omega^2 &= g_{\mu\nu} dX^\mu dX^\nu \\
&= g_{\mu\nu}^+ dX_+^\mu dX_+^\nu \sigma_+ + g_{\mu\nu}^- dX_-^\mu dX_-^\nu \sigma_+ \\
&= \left(g_{\mu\nu}^R \left[dx^\mu dx^\nu + dy^\mu dy^\nu\right] + g_{\mu\nu}^I \left[dx^\mu dy^\nu + dy^\mu dx^\nu\right]\right) \\
&\quad + I \left(g_{\mu\nu}^I \left[dx^\mu dx^\nu + dy^\mu dy^\nu\right] + g_{\mu\nu}^R \left[dx^\mu dy^\nu + dy^\mu dx^\nu\right]\right), \quad (2.52)
\end{aligned}
$$

with

$$
\begin{aligned}
g_{\mu\nu}^R &= \frac{1}{2}\left(g_{\mu\nu}^+ + g_{\mu\nu}^-\right), \\
g_{\mu\nu}^I &= \frac{1}{2}\left(g_{\mu\nu}^+ - g_{\mu\nu}^-\right). \quad (2.53)
\end{aligned}
$$

When a pseudo-real metric is used ($g_{\mu\nu}^I = 0$), this simplifies to

$$
d\omega^2 = g_{\mu\nu}^R \left[dx^\mu dx^\nu + dy^\mu dy^\nu\right] + I g_{\mu\nu}^R \left[dx^\mu dy^\nu + dy^\mu dx^\nu\right]. \quad (2.54)
$$

The first term has the same form as the length element squared proposed by M. Born
and the last term corresponds to the dispersion relation, when the dy^μ is identified
with $\left(\frac{1}{c}\right) u^\mu$. For the motion of a real particle, this term has to vanish, in order that
the $d\omega^2$ remains real.

Why to use pc-variables? The answer is that when other types of algebraic extensions are used, always ghost solutions exist [22, 23]. The way to show it is to consider
weak gravitational fields

$$
g_{\mu\nu} = \eta_{\mu\nu} + h_{\mu\nu} \quad (2.55)
$$

and to expand the Lagrangian in terms of $h_{\mu\nu}$. Field equations are obtained, which
result in operators, corresponding to unphysical particles, called ghosts and tachyons
[23]. A ghost particle has an opposite sign in the propagator as compared to physical
particles. A tachyon has an imaginary mass, which also can be seen as follows: The
mass of a relativistic moving particle is $m = \frac{m_0}{\sqrt{1-\frac{v^2}{c^2}}}$, where v is the velocity of the
particle and m_0 its rest mass. A tachyon moves at a velocity larger than c and therefore
the square root is imaginary and thus also the mass. As proved in [23], this does not
happen in the pseudo-complex algebraic extension.

Exercise 2.2 (*Ghost and tachyon solutions*)

Problem.

(a) What is the characteristic of a ghost and a tachyon solution?
(b) Show that in the limit of week gravitational fields, the pc-GR does not
    have ghost and tachyon solutions, however, a complex algebraic exten-
    sion does have at least ghost solutions. Please, consult also [23] for more
    information.

Solution.
In this exercise some knowledge of field theory is required. For details, see
[24]. Derivatives with respect to coordinates are pseudo-complex or complex
derivatives, except when stated differently.

(a) For simplicity, we restrict to real, scalar fields (bosons). The kinetic part of
the Lagrange density is

$$\pm \Phi \left(\nabla \cdot \nabla - M^2 \right) \Phi, \tag{2.56}$$

with M being the mass of the particle. The propagator is given by [24]

$$\pm \frac{1}{\left(\nabla \cdot \nabla - M^2 \right)}, \tag{2.57}$$

which in the Fourier representation and for $M = 0$ is $\pm \frac{1}{k^2}$, k being the absolute
value of the 4-wave vector.

The positive sign represents normal propagating solutions, while the negative sign corresponds to *ghost solutions*, which are not allowed to propagate
freely.

When the mass M is imaginary, one has a tachyon solution (see comments
in the last paragraph, before this exercise).

(b) For week gravitational fields one usually expands the metric up to first
order:

$$g_{\mu\nu} = \eta_{\mu\nu} + h_{\mu\nu}$$
$$g^{\mu\nu} = \eta^{\mu\nu} - h^{\mu\nu}. \tag{2.58}$$

The $\eta_{\mu\nu}$ is the metric in flat space and $h_{\mu\nu}$ describes the deviations to the flat
metric. The $g^{\mu\nu}$ is the inverse metric of $g_{\mu\nu}$ while $h^{\mu\nu}$ is the inverse of $h_{\mu\nu}$ and
the same for $\eta^{\mu\nu}$, i.e.,

$$\eta^{\mu\lambda}\eta_{\lambda\nu} = \delta_{\mu\nu}$$
$$h^{\mu\lambda}h_{\lambda\nu} = \delta_{\mu\nu}$$
$$g^{\mu\lambda}g_{\lambda\nu} = \left(\eta^{\mu\lambda} - h^{\mu\lambda}\right)\left(\eta_{\lambda\nu} + h_{\lambda\nu}\right)$$
$$\approx \left(\eta^{\mu\lambda}\eta_{\lambda\nu} + \eta^{\mu\lambda}h_{\lambda\nu} - h^{\mu\lambda}\eta_{\lambda\nu}\right)$$
$$= \delta_{\mu\nu},$$

$$(2.59)$$

where terms of second and higher order in $h_{\mu\nu}$ were neglected.

Let us take as the Lagrange density

$$\mathcal{L} = \sqrt{-g}\mathcal{R} = \sqrt{-g}g^{\mu\nu}\mathcal{R}^\alpha{}_{\mu\alpha\nu}, \qquad (2.60)$$

with $(-g)$ as the determinant of the metric, multiplied by (-1), $\mathcal{R}$ the curvature scalar and $\mathcal{R}^\alpha{}_{\mu\alpha\nu}$ the Riemann tensor, once contracted.

In the next step, we will expand $\sqrt{-g}$ up to first order, using $g^{\mu\nu}$ as given in (2.58).

The determinant $(-g)$ is given by

$$-\begin{vmatrix} -1 + h_{00} & h_{01} & h_{02} & h_{03} \\ h_{10} & 1 + h_{11} & h_{12} & h_{13} \\ h_{20} & h_{21} & 1 + h_{22} & h_{23} \\ h_{30} & h_{31} & h_{32} & 1 + h_{03} \end{vmatrix}. \qquad (2.61)$$

Expanding it up to first order gives, with $(\eta^{\mu\mu}) = (-+++)$ and noting that only the diagonal elements of the determinant contribute,

$$(-g) \approx 1 + (-h_{00} + h_{11} + h_{22} + h_{33})$$
$$= 1 + \left(\eta^{00}h_{00} + \eta^{11}h_{11} + \eta^{22}h_{22} + \eta^{33}h_{33}\right)$$
$$= 1 + h^\mu{}_\mu = 1 + h. \qquad (2.62)$$

We have defined $h = h^\mu{}_\mu$. For $\sqrt{-g}$, this gives

$$\sqrt{-g} \approx 1 + \frac{1}{2}h. \qquad (2.63)$$

The contracted Riemann tensor is given by [25, 26]

$$\mathcal{R}^\alpha{}_{\mu\alpha\nu} = \Gamma^\alpha{}_{\nu\mu,\alpha} - \Gamma^\alpha{}_{\alpha\mu,\nu} + \Gamma^\lambda{}_{\nu\mu}\Gamma^\alpha{}_{\alpha\lambda} - \Gamma^\lambda{}_{\alpha\mu}\Gamma^\alpha{}_{\nu\lambda}, \qquad (2.64)$$

where the comma "," at the end of the Γ-symbols, followed by an index μ, refers to the derivative with respect to the coordinate X^μ.

The $\Gamma^\alpha_{\beta\gamma}$ is defined as [25]

$$\Gamma^\alpha_{\beta\gamma} = \frac{1}{2}g^{\alpha\rho}\left(g_{\gamma\rho,\beta} + g_{\rho\beta,\gamma} - g_{\beta\gamma,\rho}\right)$$

$$\approx \frac{1}{2}\eta^{\alpha\rho}\left(h_{\gamma\rho,\beta} + h_{\rho\beta,\gamma} - h_{\beta\gamma,\rho}\right). \tag{2.65}$$

In the last step we expanded up to first order in $h_{\mu\nu}$ and used that deriving with respect to a coordinate only the linear term in $h_{\mu\nu}$ remains.

Substituting this result into (2.64) gives (two indices, following the comma in the symbol $h_{\mu\nu}$, refer to a double derivative)

$$\begin{aligned}
\mathscr{R}^\alpha_{\nu\alpha\mu} \approx & \frac{1}{2}\eta^{\alpha\rho}\left(h_{\mu\rho,\nu\alpha} + h_{\rho\nu,\mu\alpha} - h_{\nu\mu,\rho\alpha}\right) \\
& - \frac{1}{2}\eta^{\alpha\rho}\left(h_{\mu\rho,\alpha\nu} + h_{\rho\alpha,\mu\nu} - h_{\alpha\mu,\rho\nu}\right) \\
& + \frac{1}{4}\eta^{\lambda\rho}\eta^{\alpha\xi}\left(h_{\mu\rho,\nu} + h_{\rho\nu,\mu} - h_{\nu\mu,\rho}\right)\left(h_{\lambda\xi,\alpha} + h_{\xi\alpha,\lambda} - h_{\alpha\lambda,\xi}\right) \\
& - \frac{1}{4}\eta^{\lambda\rho}\eta^{\alpha\xi}\left(h_{\mu\rho,\alpha} + h_{\rho\alpha,\mu} - h_{\alpha\mu,\rho}\right)\left(h_{\lambda\xi,\nu} + h_{\xi\nu,\lambda} - h_{\nu\lambda,\xi}\right).
\end{aligned} \tag{2.66}$$

Using the symmetric character of $h_{\mu\nu}$ and the definitions

$$h = \eta^{\mu\lambda}h_{\lambda\mu}$$
$$A_{,\mu}^{\ \mu} = \nabla \cdot \nabla A$$
$$h_{\ \mu}^{\mu} = h_{\ \mu}^{\mu}$$

$$\text{etc.,} \tag{2.67}$$

Equation (2.66) can be written as

$$\begin{aligned}
\mathscr{R}^\alpha_{\mu\alpha\nu} \approx & \frac{1}{2}\left(h^\alpha_{\ \nu,\mu\alpha} - \nabla\cdot\nabla h_{\nu\mu} - h_{,\nu\mu} + h^\alpha_{\ \mu,\alpha\nu}\right) \\
& + \frac{1}{4}\left[\left(h^\lambda_{\ \mu,\nu} + h^\lambda_{\ \nu,\mu} - h^\lambda_{\ \nu\mu}\right)h_{,\lambda}\right. \\
& \left. - \left(h^\lambda_{\ \mu,\alpha} + h^\lambda_{\ \alpha,\mu} - h^\lambda_{\ \alpha\mu}\right)\left(h^\alpha_{\ \lambda,\nu} + h^\alpha_{\ \nu,\lambda} - h^{;\alpha}_{\ \nu\lambda}\right)\right]
\end{aligned} \tag{2.68}$$

Substituting it into the Lagrange density, gives

$$\mathscr{L} \approx \left(1 + \frac{1}{2}h\right)(\eta^{\mu\nu} - h^{\mu\nu})\mathscr{R}^\alpha_{\ \mu\alpha\nu}$$

$$\approx \left(\eta^{\mu\nu} + \frac{1}{2}h\eta^{\mu\nu} - h^{\mu\nu}\right)\mathscr{R}^\alpha_{\ \mu\alpha\nu}$$

$$\approx \left(1 + \frac{1}{2}h\right)\left(h^{\alpha\nu}_{\ ,\alpha\nu} - \nabla \cdot \nabla h\right)$$

$$- h^{\mu\nu}h^\alpha_{\ \mu,\alpha\nu} + \frac{1}{2}h^{\mu\nu}\nabla \cdot \nabla h_{\mu\nu} + \frac{1}{2}h^{\mu\nu}h_{,\mu\nu}$$

$$+ \frac{1}{4}\left[\left(2h^{\mu\lambda}_{\ ,\mu} - h^{,\lambda}\right)h_{,\lambda}\right.$$

$$\left. - \left(h^{\mu\lambda}_{\ ,\alpha} + h_\alpha^{\ \lambda,\mu} - h_\alpha^{\ \mu,\lambda}\right)\left(h^\alpha_{\ \lambda,\mu} + h^\alpha_{\ \mu,\lambda} - h^{\ \alpha}_{\mu\lambda}\right)\right], \qquad (2.69)$$

where we have used the symmetry $\eta^{\mu\nu} = \eta^{\nu\mu}$ and interchanged in some places μ with ν, after having multiplied by $\eta^{\mu\nu}$ and summed over μ and ν. In addition, only terms of up to second order in $h_{\mu\nu}$ have been taken into account.

The first line contains terms linear in h, with derivatives, e.g., $h_{,\mu}^{\ \mu}$. Using partial integration and setting surface terms to zero, in the action integral a term of the form $\int C A_{,\mu}d^4X$ changes to $-\int C_{,\mu}Ad^4X$. Because C is a constant, this term vanishes. Thus, linear terms in $h_{\mu\nu}$ do not contribute.

With this in mind, the Lagrange density changes to an effective one, given by

$$\mathscr{L} \to \mathscr{L}' = \frac{1}{2}h^{\mu\nu}\nabla \cdot \nabla h_{\mu\nu} + V(h)$$

$$V(h) = -h^{\mu\nu}h^\alpha_{\ \mu,\alpha\nu} + \frac{1}{2}h^{\mu\nu}h_{,\mu\nu}$$

$$+ \frac{1}{4}\left[\left(2h^{\mu\lambda}_{\ ,\mu} - h^{,\lambda}\right)h_{,\lambda}\right.$$

$$\left. - \left(h^{\mu\lambda}_{\ ,\alpha} + h_\alpha^{\ \lambda,\mu} - h_\alpha^{\ \mu,\lambda}\right)\left(h^\alpha_{\ \lambda,\mu} + h^\alpha_{\ \lambda,\mu} - h^{\ \alpha}_{\mu\lambda}\right)\right]. \qquad (2.70)$$

We have written explicitly the kinetic energy part and the interaction part $V(h)$ is only indicated.

Up to here, the $h_{\mu\nu}$ is either pseudo-complex or complex:

$$h_{\mu\nu} = h^R_{\mu\nu} + I h^I_{\mu\nu}$$

$$I^2 = \pm 1, \qquad (2.71)$$

where the index R refers to the (pseudo-)real and I to the (pseudo-)imaginary component. The positive sign in I^2 refers to the pseudo-complex case, while the negative sign refers to the complex case.

Considering the kinetic energy part and looking at the diagonal part only, we have

$$h^{\mu\nu}\nabla\cdot\nabla h_{\mu\nu} \rightarrow h_R^{\mu\nu}\nabla_x\cdot\nabla_x h_{\mu\nu}^R + I^2 h_I^{\mu\nu}\nabla_y\cdot\nabla_y h_{\mu\nu}^I, \qquad (2.72)$$

where we have used

$$\nabla\cdot\nabla = \frac{D^2}{DX^2}$$

$$= \left(\frac{\partial}{\partial x_\mu \partial x^\mu} + \frac{\partial}{\partial y_\mu \partial y^\mu}\right) + 2\frac{\partial}{\partial x_\mu \partial y^\mu}$$

$$\nabla_x^\mu = \frac{\partial}{\partial x_\mu}, \quad \nabla_y^\mu = \frac{\partial}{\partial y_\mu}, \qquad (2.73)$$

where x^μ and y^μ are the pseudo-real and pseudo-imaginary components of the pc-variable X^μ, and only the diagonal contributions are taken.

Inspecting (2.72), the real part of the diagonal kinetic energy has a positive sign, meaning that it yields a normal propagator. However, the diagonal part of the kinetic energy, coming from the imaginary component squared, has only a positive sign in the pseudo-complex case, i.e. no ghost propagator, while for the complex case ($I^2 = -1$) the propagator acquires a negative sign, i.e. it corresponds to a ghost propagator.

Because in both cases no mass term (with a constant M) appears, there are no tachyon solutions.

In conclusion, in the pseudo-complex case there are neither ghost nor tachyon solutions. In the complex case there are no tachyon solutions but there are ghost solutions. Thus, only the pseudo-complex case represents a consistent algebraic extension of GR.

There are other possible algebraic extensions [23], like quaternions, but all of them contain a standard complex part and, thus, ghost solutions.

In [27] a pseudo-complex field theory was proposed. The main property is that it contains a minimal length as a parameter, which is not affected by a Lorentz transformation. Due to the appearance of such a minimal length parameter, the theory is automatically regularized, i.e. there are no infinities. The pc-field Lagrange density is such that the theory is linear. The resulting propagator is the one of Pauli-Villars, which in standard field theory can only be obtained using a highly non-linear Lagrange density. Therefore, extending to pc-variables and fields the theory stays simple (linear) while the results include the minimal length. Also all continuous symmetries are maintained, which simplifies enormously the calculation of Feynman diagrams. This example proves that the use of pc-variables allows to describe systems, which are very complicated in standard theories, in an easy, elegant manner. It indicates that *nature may have a pseudo-complex structure*, as Quantum

Mechanics showed that it has a complex structure, although the final physical quantities are real.

Though, we will discuss only examples where the minimal length is set to zero, the use of pc-variables in GR may have important consequences in future: The appearance of a minimal length in pc-GR may help to quantize pc-GR!

As will be shown in the next section and in Chaps. 3 and 4 on applications, using a mapping to the four dimensional space, now physically motivated, allows an additional contribution to the Einstein equations, with the property of a repulsive energy. This energy accumulates around a large mass concentration. A collapse of a large mass is halted due to this repulsive energy, avoiding the formation of an event horizon. Though, in our "classical" theory we are only able to parametrize its contribution, demanding that no event horizon appears, the scenario is not unrealistic: As shown by Visser et al. [28–31], who discussed the Casimir effect near a mass concentration, the larger the mass, the larger the vacuum fluctuations. However, he investigated the vacuum fluctuations with a fixed back-ground metric, i.e. no back-reaction of the fluctuations to the metric was taken into account (in the absence of a quantized GR, this is a standard semi-classical treatment). Nevertheless, his work shows that the presence of mass creates, increasing with mass, vacuum fluctuations which can be associated to a dark energy. The advantage of the approach is that the density of the dark energy can be determined, however, the disadvantage is that due to the semi-classical treatment no back-reaction to the metric can be considered. Compared to this, our procedure has the advantage that a back-reaction to the metric is included, leading to a modified metric without an event horizon. However, there is also a disadvantage; Namely, that the density of the dark energy can only partly be determined from first principles: some model assumptions are necessary. As we shall see, this is rather natural.

The main point is that mass not only curves the space, as proposed in GR, but also changes the vacuum properties, implying the presence of the dark energy (vacuum fluctuations). This dark energy also distorts the space such, that space itself finally stops the collapse of a star with an arbitrary mass.

The notion of *dark energy* has to be taken here with care, because we do not know yet if it has the same origin as the cosmological dark energy, responsible for the acceleration of the universe, though we will construct such a model in Chap. 4 discussing the pc-Robertson-Walker universe.

2.6 Construction of Pseudo-complex General Relativity

The starting point is to extend the real variables of the four-dimensional space to pseudo-complex variables

$$X^\mu = X_+^\mu \sigma_+ + X_-^\mu \sigma_-, \tag{2.74}$$

which corresponds to a *product space* $W_+ \otimes W_-$, where $W_\pm$ is the manifold describing the $\sigma_\pm$ components. Because each variable has a pseudo-real and pseudo-imaginary component, one has effectively an eight-dimensional space, with the four pseudo-real and the four pseudo-imaginary components.

Due to the fact that each variable and function can be written in terms of the zero-divisor basis (see Chap. 1), we denote the metric in the form

$$g_{\mu\nu} = g_{\mu\nu}^+(X_+)\sigma_+ + g_{\mu\nu}^-(X_-)\sigma_-. \tag{2.75}$$

The two metric terms, $g_{\mu\nu}^+$ and $g_{\mu\nu}^-$, may have the same functional dependence on their corresponding variables, but even then they are different because the arguments X_+^μ and X_-^μ are different! In this chapter, we will assume the same functional dependence, i.e. $g_{\mu\nu}^+(X_+) = g_{\mu\nu}(X_+)$ and $g_{\mu\nu}^- = g_{\mu\nu}(X_-)$ and a symmetric metric, but the consideration can also be extended to a non-symmetric metric, as done by Moffat et al. [22].

In what follows, we use the same presentation as in [32–34] for the summary of geometric properties of the pc-GR: The metric is assumed to be a pseudo-holomorphic function (see Chap. 1 on the mathematics of pc-variables), i.e., it has to satisfy the pc-Riemann-Cauchy conditions

$$\frac{\partial g_{\mu\nu}^R}{\partial X_R^\lambda} = \frac{\partial g_{\mu\nu}^I}{\partial X_I^\lambda}$$
$$\frac{\partial g_{\mu\nu}^R}{\partial X_I^\lambda} = \frac{\partial g_{\mu\nu}^I}{\partial X_R^\lambda}, \tag{2.76}$$

where $g_{\mu\nu}^R$ is the pseudo-real and $g_{\mu\nu}^I$ the pseudo-imaginary component, with $X_R^\lambda = x^\lambda$ being the pseudo-real part and X_I^λ the pseudo-imaginary part of the 4-coordinate $X^\lambda = X_R^\lambda + I X_I^\lambda$.

As a distinguished property of the theory, we associate to the pseudo-imaginary component a four-velocity, i.e.,

$$X_I^\lambda = l u^\lambda. \tag{2.77}$$

where the factor l is introduced for dimensional reasons, such that X_I^λ has the dimension of length. For the special examples, discussed in this book, we set l to zero. The proposal (2.77) has its origin in the ideas of Born [4, 5] but the identification of y^μ with $l u^\mu$ has to be considered as an additional boundary condition. When we want to neglect the contributions of the minimal length, then the pseudo-imaginary term is zero and we return to the standard real coordinates x^μ. Care has to be taken with the identification (2.77). Strictly speaking, it is only valid in flat space-time and a more general identification has to be found. (See Exercise 2.4 and Chap. 7) We retain here (2.77) with the purpose to show that several formerly developed models are contained in pc-GR.

For $l > 0$, the length square element is given by

$$
\begin{aligned}
d\omega^2 &= g_{\mu\nu}(X, \mathscr{A}) DX^\mu DX^\nu \\
&= g_{\mu\nu}^+(X_+, \mathscr{A}_+) dX_+^\mu dX_+^\nu \sigma_+ + g_{\mu\nu}^-(X_-, \mathscr{A}_-) dX_-^\mu dX_-^\nu \sigma_- \\
&= g_{\mu\nu}(X, \mathscr{A}) \left(dx^\mu dx^\nu + l^2 du^\mu du^\nu \right) + g_{\mu\nu}(X, \mathscr{A}) 2lI dx^\mu du^\nu.
\end{aligned}
$$

(2.78)

We have used the abbreviation $\mathscr{A}$ for all parameters of the theory. Note, that the (symmetric) metric depends on the coordinates and the parameters only.

Because the zero-divisor components act independently, it is allowed to use the same definitions and methods as given in detail in [25, 35]. For example, the parallel displacement of a pc-vector is given by

$$
\begin{aligned}
DW^\mu &= -\Gamma_{\nu\lambda}^\mu DX^\nu W^\lambda \\
&= -\Gamma_{\nu\lambda}^{+\,\mu} DX_+^\nu W_+^\lambda \sigma_+ - \Gamma_{\nu\lambda}^{-\,\mu} DX_-^\nu W_-^\lambda \sigma_- \\
&= dW_+^\mu \sigma_+ + dW_-^\mu \sigma_-,
\end{aligned}
$$

(2.79)

where DX^ν refers to the change of the pseudo-complex coordinate X^ν and W^μ are the components of a vector. For the moment, we exclude torsion and therefore the connections $\Gamma_{\nu\lambda}^\mu$ are symmetric in their lower indices (Levi-Civita connection [25]).

The pc-Christoffel symbols are written in terms of the zero-divisor basis as

$$
\begin{aligned}
\Gamma_{\mu\nu}^\lambda &= \left\{ \begin{matrix} \lambda \\ \nu\mu \end{matrix} \right\} \\
&= \left\{ \begin{matrix} \lambda \\ \nu\mu \end{matrix} \right\}_+ \sigma_+ + \left\{ \begin{matrix} \lambda \\ \nu\mu \end{matrix} \right\}_- \sigma_-,
\end{aligned}
$$

(2.80)

which are expressed in terms of *Christoffel symbols of the second kind* [25] as

$$
\Gamma_{\mu\nu}^{\pm\,\lambda} = \left\{ \begin{matrix} \lambda \\ \nu\mu \end{matrix} \right\}_\pm = g_\pm^{\lambda\kappa} [\nu\mu, \kappa]_\pm .
$$

(2.81)

The *Christoffel symbols of the first kind* are defined as in [25]

$$
[\mu\nu, \kappa] = \frac{1}{2} \left(\frac{Dg_{\mu\kappa}}{DX^\nu} + \frac{Dg_{\nu\kappa}}{DX^\mu} - \frac{Dg_{\mu\nu}}{DX^\kappa} \right).
$$

(2.82)

(In general, torsion can be included, leading to additional terms in (2.81).)

From (2.80) the real and pseudo-imaginary components are deduced:

$$\left\{ \begin{matrix} \lambda \\ v\mu \end{matrix} \right\}_R = \frac{1}{2} \left(\left\{ \begin{matrix} \lambda \\ v\mu \end{matrix} \right\}_+ + \left\{ \begin{matrix} \lambda \\ v\mu \end{matrix} \right\}_- \right)$$

$$\left\{ \begin{matrix} \lambda \\ v\mu \end{matrix} \right\}_I = \frac{1}{2} \left(\left\{ \begin{matrix} \lambda \\ v\mu \end{matrix} \right\}_+ - \left\{ \begin{matrix} \lambda \\ v\mu \end{matrix} \right\}_- \right). \tag{2.83}$$

The 4-derivative of a contravariant vector is given by

$$W_{||v}^{\mu} = W_{|v}^{\mu} + \left\{ \begin{matrix} \mu \\ v\lambda \end{matrix} \right\} W^{\lambda}$$

$$= \left(W_{+|v}^{\mu} + \left\{ \begin{matrix} \mu \\ v\lambda \end{matrix} \right\}_+ W_+^{\lambda} \right) \sigma_+$$

$$+ \left(W_{-|v}^{\mu} + \left\{ \begin{matrix} \mu \\ v\lambda \end{matrix} \right\}_- W_-^{\lambda} \right) \sigma_-, \tag{2.84}$$

where $W_{|v}^{\mu} = \frac{DW^{\mu}}{DX^v}$. The rules for deriving covariant vectors and tensors can be directly copied from [25] (with the change in the signature of the metric).

An important point is that in this new formulation the 4-derivative of the metric will again be zero. To show this, we copy the arguments, as given in [25], Chap. 3. We have

$$g_{\mu v|\lambda}^{\pm} - g_{\mu\kappa}^{\pm} \left\{ \begin{matrix} \kappa \\ v\lambda \end{matrix} \right\}_{\pm} = [\mu\lambda, v]_{\pm}, \tag{2.85}$$

where the symmetry property of the metric tensor was used. Equation (2.85) is proved by substituting the Christoffel symbol of the second kind (2.81) and using the definition of the Christoffel symbol of the first kind (2.82).

The divergence of $g_{\mu v}^{\pm}$ is given by

$$g_{\mu v||\lambda}^{\pm} = g_{\mu v|\lambda}^{\pm} - \left\{ \begin{matrix} \kappa \\ v\lambda \end{matrix} \right\}_{\pm} g_{\mu\kappa}^{\pm} - \left\{ \begin{matrix} \kappa \\ \mu\lambda \end{matrix} \right\}_{\pm} g_{\kappa v}^{\pm}$$

$$= [\mu\lambda, v]_{\pm} - g_{\kappa v}^{\pm} \left\{ \begin{matrix} \kappa \\ \mu\lambda \end{matrix} \right\}_{\pm} \tag{2.86}$$

The first line is nothing but the covariant transformation of a tensor with two indices [25], while in the second line (2.85) was substituted.

Utilizing the definition of the Christoffel symbol of the second kind, this expression is identical to zero. Thus, also the 4-derivative of the pseudo-complex metric is zero:

$$g_{\mu v||\lambda} = g_{\mu v||\lambda}^{+}\sigma_+ + g_{\mu v||\lambda}^{-}\sigma_- = 0, \tag{2.87}$$

or equivalently

$$g^{\pm}_{\mu\nu||\lambda} = 0, \tag{2.88}$$

where the derivative is with respect to the coordinates $X^{\lambda}_{\pm}$.

This result is very important because it states that a *universal pc-metric*, namely a metric which is invariant under the transformation to another system, can be defined.

Exercise 2.3 (*A relation involving the covariant derivative of the metric*)

Problem. Prove (2.85), which is used in (2.86).

Solution. The first line in (2.86) is nothing but the covariant derivative of a second rank tensor with two covariant indices [25]. What has to be shown is that the first two terms are expressed with the Christoffel symbols of the first kind (see (2.85)). The prove is taken from [25].

We skip for the moment the $\pm$-indices, because what will be shown is equivalent for each zero-divisor component.

Let us start, considering a vector field $W^{\nu}(s)$ at each point of a curve satisfying the equation

$$\frac{dW^{\nu}}{ds} + \left\{ \begin{matrix} \nu \\ \xi\lambda \end{matrix} \right\} \frac{dx^{\lambda}}{ds} W^{\xi} = 0. \tag{2.89}$$

These are ordinary first-order differential equations and the $W^{\nu}(s)$ are defined along the curve, once initial values have been chosen.

Consider the quantity

$$P(s) = g_{\mu\nu} V^{\mu} W^{\nu}, \tag{2.90}$$

where W^{ν} the contra-variant components of an arbitrary 4-vector and s the curve parameter. $P(s)$, defined for each point of a curve is clearly a scalar. This implies that

$$\frac{dP(s)}{ds} = P'(s) \tag{2.91}$$

is also a scalar. Written explicitly gives, using the product and chain rule for differentiation:

$$P'(s) = g_{\mu\nu|\lambda} \frac{dx^{\lambda}}{ds} V^{\mu} W^{\nu} + g_{\mu\nu} V^{\mu}_{|\lambda} \frac{dx^{\lambda}}{ds} W^{\nu} + g_{\mu\nu} V^{\mu} W^{\nu}_{|\lambda} \frac{dx^{\lambda}}{ds}. \tag{2.92}$$

On the right hand side of (2.92) we relabel in the first and second term v to ξ and in the last term we substitute (2.89). This gives

$$P'(s) = g_{\mu\xi|\lambda}\frac{dx^\lambda}{ds}V^\mu W^\xi + g_{\mu\xi}V^\mu_{|\lambda}\frac{dx^\lambda}{ds}W^\xi - g_{\mu\nu}V^\mu \left\{\begin{matrix} v \\ \xi\lambda \end{matrix}\right\}\frac{dx^\lambda}{ds}W^\xi$$

$$= \left[V^\mu\left(g_{\mu\xi|\lambda} - g_{\mu\nu}\left\{\begin{matrix} v \\ \xi\lambda \end{matrix}\right\}\right) + g_{\mu\xi}V^\mu_{|\lambda}\right]\frac{dx^\lambda}{ds}W^\xi$$

$$= T_{\xi\lambda}\frac{dx^\lambda}{ds}W^\xi, \tag{2.93}$$

where the matrix $T_{\xi\lambda}$ has been defined. The left hand side of this equation is a scalar and $\frac{dx^\lambda}{ds}$, W^ξ are arbitrary vectors at x^λ. Therefore, by the quotient theorem [25] the $T_{\xi\lambda}$ has to be a tensor. The form of this tensor can be simplified using the definitions of the Christoffel symbols of the first and second kind, whose relation is given in (2.81). One obtains finally

$$g_{\mu\xi|\lambda} - g_{\mu\nu}\left\{\begin{matrix} v \\ \xi\lambda \end{matrix}\right\} = g_{\mu\xi|\lambda} - g_{\mu\nu}g^{\nu\kappa}[\xi\lambda,\kappa]$$

$$= g_{\mu\xi|\lambda} - \delta_{\mu\kappa}[\xi\lambda,\kappa]$$

$$= g_{\mu\xi|\lambda} - [\xi\lambda,\mu]$$

$$= g_{\mu\xi|\lambda} - \frac{1}{2}\left(g_{\xi\mu|\lambda} + g_{\lambda\mu|\xi} - g_{\xi\lambda|\mu}\right)$$

$$= \frac{1}{2}\left(g_{\mu\xi|\lambda} + g_{\lambda\xi|\mu} - g_{\lambda\mu|\xi}\right)$$

$$= [\mu\lambda,\xi], \tag{2.94}$$

which proves relation (2.85) and thus (2.86), one has only to add the indices $\pm$. In (2.94) the symmetric nature of the metric tensor was used.

Finally, the expressions for the Riemann scalar, Ricci tensor and Riemann tensor can be readily copied from [25, 35], with the difference that the Christoffel symbols are substituted by their pc-counterparts and a derivative by a pc-derivative.

In this manner, the formulation of pc-GR is very similar to the standard GR.

2.7 A Modification of the Variational Principle

In [36, 37] a modified variational principle was proposed, such that the pc-extension of a theory produces a new theory. The argument goes as follow: When the variation of the action

$$S = \int L \, d^4x \qquad (2.95)$$

is set to zero, one obtains

$$\delta S_\pm = 0, \qquad (2.96)$$

because the $\sigma_\pm$ components separate and do not mix. The two independent components can be treated as two different theories and nothing new is obtained. Therefore, [36, 37] proposes to modify the metric such that

$$\delta S = \delta S_+ \sigma_+ + \delta S_- \sigma_- \in \text{ zero divisor.} \qquad (2.97)$$

The element in the zero divisor is either proportional to σ_+ or to σ_- and has zero norm, as was shown in Chap. 1. Therefore, it is allowed to call this object a *generalized zero*. As an example, let us assume that the variation of S is proportional to σ_-, i.e. $\lambda\sigma_-$. The following discussion is completely the same when σ_+ is used instead. The variation of the action gives the equations

$$\mathbf{F}_+ \sigma_+ + \mathbf{F}_- \sigma_- = \lambda\sigma_-, \qquad (2.98)$$

with $\mathbf{F}_\pm$ some particular differential operators, depending on S. In [36, 37] one proceeds by multiplying (2.98) with its pseudo-complex conjugate, using $\sigma_+ \sigma_- = 0$, $\sigma_\pm^2 = \sigma_\pm$ and $(\sigma_+ + \sigma_-) = 1$. The results is

$$
\begin{aligned}
(F_+\sigma_+ + F_-\sigma_-)\,(F_+\sigma_+ + F_-\sigma_-)^* \\
= (F_+\sigma_+ + F_-\sigma_-)\,(F_+\sigma_- + F_-\sigma_+) \\
= F_+ F_-\,(\sigma_+ + \sigma_-) \\
= F_+ F_- = \lambda\sigma_-\sigma_+ = 0.
\end{aligned}
\qquad (2.99)
$$

which appears like a usual equation of motion. However, while $F_\pm$ are linear operators, the product is a highly non-linear operator.

An alternative, but equivalent way is to start from (2.98) and to exploit the linear independence of the zero-divisor components. This leads to

$$F_+ = 0 \quad \text{and} \quad F_- = \lambda. \qquad (2.100)$$

It is not difficult to repeat these steps in order to obtain the equations when the element in the zero-divisor is proportional to σ_+. The formulation is symmetrical.

In the case of pc-GR, the $\sigma_\pm$ components of the Einstein tensor are

$$G_{\mu\nu}^\pm = \mathscr{R}_{\mu\nu}^\pm - \frac{1}{2} g_{\mu\nu}^\pm \mathscr{R}^\pm, \qquad (2.101)$$

with $\mathscr{R}^{\pm}_{\mu\nu}$ the Ricci tensor in each zero-divisor component and $\mathscr{R}^{\pm}$ is the curvature scalar. The $g^{\pm}_{\mu\nu}$ is the metric in each component.

Exercise 2.4 (*An alternative justification for the modified variational principle*)

The objective of this to give a foundation of the modified variational principle. For that, one has to show how to implement the constraint that the pseudo-imaginary component of the pc-length element squared, $d\omega_I^2$, is zero. This component is the difference of the σ_+ to the σ_- component, thus,

$$g^{-}_{\mu\nu}\dot{X}^{\mu}_{-}\dot{X}^{\nu}_{-} = g^{+}_{\mu\nu}\dot{X}^{\mu}_{+}\dot{X}^{\nu}_{+}, \tag{2.102}$$

with $\dot{X}^{\mu}_{\pm} = \frac{dX^{\mu}_{\pm}}{ds}$, i.e., one assumes that $s_{\pm} = s$ the same in both zero-divisor components. Here, when s is the proper time $-\tau$, one assumes that $d\tau_+ = d\tau_-$.

The method to use is the one of *Lagrange multipliers*, which is well explained in any book on Classical Mechanics. The main idea is, that due to a constraint (let us restrict to only one), not all variables, which are varied, are linear independent. In case of one constraint expressed only in the variable $f(g_{\mu\nu}) = 0$, one variable can be substituted by the other ones. However, this not always gives an easy expression and it is recommended to use Lagrange multipliers.

In GR (including pc-GR), the variables to vary are $g_{\mu\nu}$ (in fact, one has to bear in mind that due to the symmetry properties there are only $4(4-1)/2 = 6$ independent variables ("coordinates")). In pc-GR we have the variables $g^{\pm}_{\mu\nu}$.

The variation with respect to the metric is applied to the action:

$$\frac{\delta S}{\delta g_{\mu\nu}}\delta g_{\mu\nu} = \frac{\delta S_+}{\delta g^{+}_{\mu\nu}}\delta g^{+}_{\mu\nu}\sigma_+ + \frac{\delta S_-}{\delta g^{-}_{\mu\nu}}\delta g^{-}_{\mu\nu}\sigma_- = 0. \tag{2.103}$$

Note, that it is not set to a zero-divisor, because we want to show that the additional contribution is the consequence of the constraint (2.102). Note also, that in the σ_+ sector, *only a variation with respect to $g^{+}_{\mu\nu}$ is applied*, while in the σ_- sector *the variation is only with respect to $g^{-}_{\mu\nu}$*! Thus, adding a function which also depends on $g^{+}_{\mu\nu}$ in the σ_- sector, the $g^{+}_{\mu\nu}$ has to be treated as a constant.

Applying the variation independently in the two zero-divisor sectors, leads to

$$\int \left[\mathscr{R}^{\mu\nu}_{\pm} - \frac{1}{2}g^{\mu\nu}_{\pm}\mathscr{R}_{\pm}\right]\delta g^{\pm}_{\mu\nu}ds = 0 \tag{2.104}$$

One can not set the expressions in the square bracket to zero, because the $\delta g_{\mu\nu}^{\pm}$ are not linearly independent. ($\mu \leq \nu$).

To overcome this situation, the constraint (2.102) is varied with respect to $g_{\mu\nu}^{-}$ and added to the σ_- component. One can do it also the other way around and vary $g_{\mu\nu}^{+}$ and add it to the σ_+ component. This is just a matter of definition where to add the constraint. One considers the $\dot{X}^{\mu}_{-}$ as *arbitrary put fixed components of a vector in the tangent space*. These are, thus, not considered as the variables to vary with.

The variation of the constraint is

$$\frac{\delta \left(g_{\lambda\rho}^{+} \dot{X}_{+}^{\lambda} \dot{X}_{+}^{\rho} - g_{\lambda\rho}^{-} \dot{X}_{-}^{\lambda} \dot{X}_{-}^{\rho}\right)}{\delta g_{\mu\nu}^{-}} \delta g_{\mu\nu}^{-} = -\left(\dot{X}_{-}^{\mu} \dot{X}_{-}^{\nu}\right) \delta g_{\mu\nu}^{-}. \quad (2.105)$$

Adding this to (2.104) and considering only the σ_- component, because only there the constraint is added and therefore changes the expression, we obtain

$$\int \left[\mathcal{R}_{-}^{\mu\nu} - \frac{1}{2} g_{-}^{\mu\nu} \mathcal{R}_{-} - \lambda \left(\dot{X}_{-}^{\mu} \dot{X}_{-}^{\nu}\right) \right] \delta g_{\mu\nu}^{-} ds = 0, \quad (2.106)$$

where λ is the *Lagrange multiplyer*.

The λ is chosen such that the term corresponding to the linear dependent variable, say g_{00}^{-} vanishes. This gives one equation. Then only terms multiplied with a linear independent differential $\delta g_{\mu\nu}^{-}$ appear and the expressions in the square parenthesis can all be set to zero. This gives $(6 - 1) = 5$ equations. Note, that one has not to do the explicit calculation, because all what matters here is that it can be done. At the end one has 6 equations which appear equal.

This leads to the following equations

$$\mathcal{R}_{-}^{\mu\nu} - \frac{1}{2} g_{-}^{\mu\nu} \mathcal{R}_{-} - \lambda \left(\dot{X}_{-}^{\mu} \dot{X}_{-}^{\nu}\right) = 0$$

$$g_{\mu\nu}^{-} \dot{X}_{-}^{\mu} \dot{X}_{-}^{\nu} = g_{\mu\nu}^{+} \dot{X}_{+}^{\mu} \dot{X}_{+}^{\nu}. \quad (2.107)$$

One has to pay the price to have in addition of the equations, coming from the variation, also the constraint equation. All equations in (2.107) together have to be solved.

Lowering all indices in the first equation in (2.107) and using

$$\dot{X}_{\mu}^{-} = \dot{x}_{\mu} - \dot{y}_{\mu} = u_{\mu} - \dot{y}_{\mu}, \quad (2.108)$$

we can rewrite it as follows

$$\mathscr{R}^-_{\mu\nu} - \frac{1}{2}g^-_{\mu\nu}\mathscr{R}_- = \lambda \left(\dot{X}^-_\mu \dot{X}^-_\nu\right)$$

$$= \lambda u_\mu u_\nu + \lambda \left(\dot{y}_\mu \dot{y}_\nu - u_\mu \dot{y}_\nu - u_\nu \dot{y}_\mu\right). \quad (2.109)$$

Care has to be taken in raising and lowering indices. One can not do it individually but only in combinations, i.e. only the upper index of X^μ_- can be lowered by $g^-_{\mu\nu}$ and the lower index X^-_μ can be raised by $g^{\mu\nu}_-$.

The right hand side of the Einstein equation can be written in terms of a tensor which has to be identified with an energy-momentum tensor, multiplied with $-\frac{8\pi\kappa}{c^2}$. For that, we redefine the Lagrange multiplier as

$$\lambda = \frac{8\pi\kappa}{c^2}\tilde{\lambda}. \quad (2.110)$$

There are two expressions

$$\left(T^\Lambda\right)_{\mu\nu} = \tilde{\lambda}\left[u_\mu u_\nu + \dot{y}_\mu \dot{y}_\nu - u_\mu \dot{y}_\nu - u_\nu \dot{y}_\mu\right]$$
$$\rightarrow$$
$$\left(T^{\text{fluid}}\right)_{\mu\nu} = \left(\varepsilon_\Lambda + \frac{p_\Lambda}{c^2}\right)u_\mu u_\nu + \frac{p}{c^2}g_{\mu\nu}, \quad (2.111)$$

where the last row refers to a mapping to make, in case the energy-momentum tensor is identified as a fluid. The above expression permits also other interpretations.

Here, one has to make a parenthesis: A fluid is a *macroscopic* concept, where granular, microscopic structures are averaged out. Thus, also here the granular structure due to the minimal length is averaged out and the minimal length does not appear any more. Also, the metric which has to be used is the projected, approximated metric $(g_{\mu\nu})$, which does not depend any more on the minimal length.

This discussion on constraints not only gives the foundation for the modified variational principle, but also shows the symmetry between the σ_+ and σ_- component, i.e. that the same can be done by adding the constraint equation to the σ_+ component. It also gives a hint on how to obtain y^μ.

Using the modified variational procedure, leads to an additional contribution $\lambda\sigma_-$ (having used the convention that only elements in the σ_- direction are used). The λ must be a tensor with the indices $(\mu\nu)$, which can be rewritten in terms of $-\frac{8\pi\kappa}{c^2}\left(T^\Lambda\right)_{\mu\nu}$, κ being the gravitational constant, c the velocity of light and Λ points to its character as a dark energy. Thus, the modified Einstein equations are

$$G^+_{\mu\nu} = 0 \quad \text{and} \quad G^-_{\mu\nu} = -\frac{8\pi\kappa}{c^2}\left(T^\Lambda\right)_{\mu\nu}, \quad (2.112)$$

where $\left(T^{\Lambda}\right)_{\mu\nu}$ is an energy-momentum tensor. We will see in the pc-Schwarzschild and pc-Kerr solutions (Chap. 3) that this energy is repulsive and provides the possibility to eliminate the event horizon. Thus, the massive object is not a black hole but rather a *dark star* (the notation is loaned from [28]).

The basic concept behind this is the one explained in the preface, namely that a mass not only curves the space, as in GR, but also modifies the local structure of space around the mass.

To solve (2.112), in order to obtain the metric, one has to solve each zero divisor component separately. Because the σ_+ component obeys the same rules as standard GR, the solutions (Schwarzschild, Kerr, etc.) are the same, however the σ_- components has to be solved in the presence of an additional energy momentum tensor. It is similar to solve GR in the presence of an energy distribution. The final result leads to an apparently different function $g_{\mu\nu}^{\pm}$.

The problem is, once the pc-metric is obtained, how we map to the physical four-dimensional space? We will justify a possibility, discussing as an example the group theoretical structure of the pc-Lorentz group. It *leads to the rule* for the mapping to the physical space.

The generators of the pc-Lorentz group are given by the antisymmetric operators

$$
\begin{aligned}
\mathcal{L}_{\mu\nu} &= X_\mu P_\nu - X_\nu P_\nu \\
&= \left(X_\mu^+ P_\nu^+ - X_\nu^+ P_\mu^+\right)\sigma_+ + \left(X_\mu^- P_\nu^- - X_\nu^- P_\mu^-\right)\sigma_- \\
&= L_{\mu\nu}^+\sigma_+ + L_{\mu\nu}^-\sigma_-,
\end{aligned}
\tag{2.113}
$$

where X_μ and P_ν are pc-coordinates and -momenta. The division into the zero-divisor components is also given and $L_{\mu\nu}^{\pm}$ are the corresponding generators of the Lorentz transformation in each sector.

A finite pc-Lorentz transformation is

$$
e^{-i\omega^{\mu\nu}\mathcal{L}_{\mu\nu}} = e^{-i\omega_+^{\mu\nu}\mathcal{L}_{\mu\nu}^+}\sigma_+ + e^{-i\omega_-^{\mu\nu}\mathcal{L}_{\mu\nu}^-}\sigma_-,
\tag{2.114}
$$

having in each zero divisor component a Lorentz transformation and thus a Lorentz group $SO_\pm(3, 1)$. The parameters of the transformation are also pseudo-complex and have the structure

$$
\omega^{\mu\nu} = \omega_+^{\mu\nu}\sigma_+ + \omega_-^{\mu\nu}\sigma_-.
\tag{2.115}
$$

Because the expression in the σ_+ sector commutes with the one in the σ_- sector (remember that $\sigma_+\sigma_- = 0$), the group structure is given by

$$
SO_{\mathscr{P}}(3, 1) = SO_+(3, 1) \otimes SO_-(3, 1) \supset SO(3, 1).
\tag{2.116}
$$

The pc-Lorentz group is therefore a direct product of two independent Lorentz groups.

This can be reduced to the usual Lorentz group by setting $\omega_+ = \omega_-$ and $l = 0$, which implies that $X_\mu \to x_\mu$ and $P_\nu \to p_\nu$. Setting $l = 0$ is equivalent to neglecting the effects of a maximal acceleration (in a previous section we showed that the appearance of a minimal length implies a maximal acceleration of $\frac{1}{l^2}$, i.e., when l is set to zero, there is no limit any more). Another way to put it is to set the pseudo-imaginary component of X^μ, which is given by lu^μ, to zero. Therefore, one can state that *the reduction to the standard Lorentz group is achieved by substituting all parameters and variables in the theory by their pseudo-real components, i.e. setting all pseudo-imaginary components to zero.*

Exercise 2.5 (*Properties of the pc-Lorentz transformation I*)

Problem.
(a) The infinitesimal generators of the pseudo-complex Lorentz transformation are

$$\mathcal{L}_{\mu\nu} = X_\mu P_\nu - X_\nu P_\mu$$

with

$$P_\mu = \frac{\hbar}{i}\frac{D}{DX^\mu}.$$

Determine the commutation relations.
(b) Map $\mathcal{L}_{\mu\nu}$ to

$$L_{\mu\nu} = x_\mu p_\nu - x_\nu p_\mu$$

with

$$p_\mu = \frac{\hbar}{i}\frac{\partial}{\partial x^\mu} \tag{2.117}$$

and consider the finite transformation

$$\Lambda = e^{i\omega^{\mu\nu}L_{\mu\nu}},$$

with $\omega_{\mu\nu}$ as the pseudo-complex transformation parameters.
Show that Λ can be written as a standard Lorentz transformation, with purely real parameters, and a transformation with purely pseudo-complex parameters.

Solution. (a) The commutator of two generators is

$$\left[\mathscr{L}_{\mu\nu}, \mathscr{L}_{\lambda\delta}\right] = \left[X_\mu P_\nu - X_\nu P_\mu, X_\lambda P_\delta - X_\delta P_\lambda\right]. \qquad (2.118)$$

The commutator between X^μ and P_ν is given by

$$\left[X^\mu, P_\nu\right] = \left[X^\mu, \frac{\hbar}{i}\frac{D}{DX^\nu}\right] = i\hbar\delta_{\mu\nu}. \qquad (2.119)$$

This implies that

$$\left[X_\mu, P_\nu\right] = g_{\mu\eta}\left[X^\eta, P_\nu\right] = i\hbar g_{\mu\eta}\delta_{\eta\nu} = i\hbar g_{\mu\nu}. \qquad (2.120)$$

The appearance of the metric component is due to the fact that the metric is non-euclidean. In an euclidean space the $g_{\mu\nu}$ is just $-\delta_{\mu\nu}$.

With this, the commutation relations give

$$
\begin{aligned}
&\left[X_\mu P_\nu, X_\lambda P_\delta\right] - \left[X_\mu P_\nu, X_\delta P_\lambda\right] - \left[X_\nu P_\mu, X_\lambda P_\delta\right] + \left[X_\nu P_\mu, X_\delta P_\lambda\right] \\
&= X_\mu\left[P_\nu, X_\lambda\right]P_\delta + X_\lambda\left[X_\mu, P_\delta\right]P_\nu \\
&\quad - X_\mu\left[P_\nu, X_\delta\right]P_\lambda - X_\delta\left[X_\mu, P_\lambda\right]P_\nu \\
&\quad - X_\nu\left[P_\mu, X_\lambda\right]P_\delta - X_\lambda\left[X_\nu, P_\delta\right]P_\mu \\
&\quad + X_\nu\left[P_\mu, X_\delta\right]P_\lambda + X_\delta\left[X_\nu, P_\lambda\right]P_\mu \\
&= -i\hbar\left(g_{\nu\lambda}X_\mu P_\delta - g_{\mu\delta}X_\lambda P_\nu - g_{\nu\delta}X_\mu P_\lambda + g_{\mu\lambda}X_\delta P_\nu\right. \\
&\quad \left. + g_{\mu\lambda}X_\nu P_\delta + g_{\nu\delta}X_\lambda P_\mu + g_{\mu\delta}X_\nu P_\lambda - g_{\nu\lambda}X_\delta P_\mu\right) \\
&= i\hbar\left(g_{\lambda\nu}\left(X_\delta P_\mu - X_\mu P_\delta\right) + g_{\delta\mu}\left(X_\lambda P_\nu - X_\nu P_\lambda\right)\right. \\
&\quad \left. + g_{\lambda\mu}\left(X_\nu P_\delta - X_\delta P_\nu\right) + g_{\delta\nu}\left(X_\mu P_\lambda - X_\lambda P_\mu\right)\right).
\end{aligned}
$$
$$(2.121)$$

On the right hand side appear again generators of the pc-Lorentz group. With this, the final result for the commutation relation of the generators is

$$\left[\mathscr{L}_{\mu\nu}, \mathscr{L}_{\lambda\delta}\right] = i\hbar\left(g_{\lambda\nu}\mathscr{L}_{\delta\mu} + g_{\delta\mu}\mathscr{L}_{\lambda\nu} + g_{\lambda\mu}\mathscr{L}_{\nu\delta} + g_{\delta\nu}\mathscr{L}_{\mu\lambda}\right). \qquad (2.122)$$

(b) We use the mathematical properties of pseudo-complex variables and functions, as explained in Chap. 1.

The finite Lorentz transformation, with $L_{\mu\nu}$ as the generators is written in terms of pc-transformation parameters $\omega^{\mu\nu}$ as follows:

$$
\begin{aligned}
\Lambda = e^{\omega^{\mu\nu}L_{\mu\nu}} &= e^{(\omega_R^{\mu\nu} + I\omega_I^{\mu\nu})L_{\mu\nu}} = e^{\omega_+^{\mu\nu}L_{\mu\nu}\sigma_+ + \omega_-^{\mu\nu}L_{\mu\nu}\sigma_-} \\
&= e^{\omega_+^{\mu\nu}L_{\mu\nu}}\sigma_+ + e^{\omega_-^{\mu\nu}L_{\mu\nu}}\sigma_- \\
&= \Lambda_+\sigma_+ + \Lambda_-\sigma_-.
\end{aligned}
$$
$$(2.123)$$

Using $\omega_{\pm}^{\mu\nu} = \omega_R^{\mu\nu} \pm \omega_I^{\mu\nu}$ we can write this transformation as

$$
\begin{aligned}
\Lambda &= e^{\left(\omega_R^{\mu\nu}+\omega_I^{\mu\nu}\right)L_{\mu\nu}}\sigma_+ + e^{\left(\omega_R^{\mu\nu}-\omega_I^{\mu\nu}\right)L_{\mu\nu}}\sigma_- \\
&= \left(e^{\omega_I^{\mu\nu}}\sigma_+ + e^{-\omega_I^{\mu\nu}}\sigma_-\right) \\
&\quad \left(e^{\omega_R^{\mu\nu}}\sigma_+ + e^{+\omega_R^{\mu\nu}}\sigma_-\right) \\
&= \left(\Lambda_+^{(2)}\sigma_+ + \Lambda_-^{(2)}\sigma_-\right)\left(\Lambda_+^{(1)}\sigma_+ + \Lambda_-^{(1)}\sigma_-\right) \\
&= \Lambda^{(2)}\Lambda^{(1)},
\end{aligned}
\tag{2.124}
$$

where the real proper Lorentz transformations $\Lambda^{(k)}$ ($k = 1, 2$) have been introduced.

The following properties can be deduced from (2.124):

$$
\Lambda_-^{(1)} = \Lambda_+^{(1)}, \ \Lambda_-^{(2)} = \left(\Lambda_+^{(2)}\right)^{-1},
\tag{2.125}
$$

because

$$
\begin{aligned}
\Lambda^{(1)} &= e^{\omega_R^{\mu\nu}L_{\mu\nu}}\sigma_+ + e^{\omega_R^{\mu\nu}L_{\mu\nu}}\sigma_- = e^{\omega_R^{\mu\nu}L_{\mu\nu}} \\
\Lambda^{(2)} &= e^{\omega_I^{\mu\nu}L_{\mu\nu}}\sigma_+ - e^{\omega_I^{\mu\nu}L_{\mu\nu}}\sigma_- = e^{I\omega_I^{\mu\nu}L_{\mu\nu}}.
\end{aligned}
\tag{2.126}
$$

The last equations are the required answer of the problem.

Exercise 2.6 (*Properties of the pc-Lorentz transformation II*)

Problem.
Show that the pc-Lorentz transformation

$$
\Lambda = e^{i\omega_{\mu\nu}L_{\mu\nu}}
$$

implies a maximal acceleration.

Hint: Consider a boost in the x-direction and use the results of Exercise 2.5.

Solution. The basis is the solution of the Lorentz transformation as given in Problem No. 1. A boost in the x direction for $\Lambda_+^{(2)}$ is given by

$$\begin{pmatrix} \gamma & -\beta\gamma & 0 & 0 \\ -\beta\gamma & \gamma & 0 & 0 \\ 0 & 0 & 0 & 0 \\ 0 & 0 & 0 & 1 \end{pmatrix},$$ (2.127)

with $\beta = \frac{v}{c}$ and $\gamma = \sqrt{\left(1 - \frac{v^2}{c^2}\right)}$.

Define the rapidity ϕ via

$$\gamma = \cosh\phi$$
$$\beta\gamma = \sinh\phi,$$ (2.128)

from which follows

$$\beta = \tanh\phi.$$ (2.129)

The $\Lambda_+^{(2)}$ is then given by

$$\Lambda_+^{(2)} = \begin{pmatrix} \cosh\phi & -\sinh\phi & 0 & 0 \\ -\sinh\phi & \cosh\phi & 0 & 0 \\ 0 & 0 & 1 & 0 \\ 0 & 0 & 0 & 1 \end{pmatrix}.$$ (2.130)

The inverse of this matrix is

$$\left(\Lambda_+^{(2)}\right)^{-1} = \begin{pmatrix} \cosh\phi & \sinh\phi & 0 & 0 \\ \sinh\phi & \cosh\phi & 0 & 0 \\ 0 & 0 & 1 & 0 \\ 0 & 0 & 0 & 1 \end{pmatrix},$$ (2.131)

which is verified easily.

Using, that the transformation $\Lambda^{(2)}$ is given by

$$\Lambda^{(2)} = \frac{1}{2}\left(\Lambda_+^{(2)} + \left(\Lambda_+^{(2)}\right)^{-1}\right) + \frac{I}{2}\left(\Lambda_+^{(2)} - \left(\Lambda_+^{(2)}\right)^{-1}\right)$$
$$= \Lambda_R^{(2)} + I\Lambda_I^{(2)}$$ (2.132)

and using (2.130), (2.131), we obtain for the pseudo-real and pseudo-imaginary component

$$\Lambda_R^{(2)} = \frac{1}{2} \left[\begin{pmatrix} \cosh\phi & -\sinh\phi & 0 & 0 \\ -\sinh\phi & \cosh\phi & 0 & 0 \\ 0 & 0 & 1 & 0 \\ 0 & 0 & 0 & 1 \end{pmatrix} + \begin{pmatrix} \cosh\phi & \sinh\phi & 0 & 0 \\ \sinh\phi & \cosh\phi & 0 & 0 \\ 0 & 0 & 1 & 0 \\ 0 & 0 & 0 & 1 \end{pmatrix} \right]$$

$$= \frac{1}{2} \begin{pmatrix} 2\cosh\phi & 0 & 0 & 0 \\ 0 & 2\cosh\phi & 0 & 0 \\ 0 & 0 & 1 & 0 \\ 0 & 0 & 0 & 1 \end{pmatrix}$$

$$= \cosh\phi \begin{pmatrix} 1 & 0 & 0 & 0 \\ 0 & 1 & 0 & 0 \\ 0 & 0 & 1 & 0 \\ 0 & 0 & 0 & 1 \end{pmatrix} \tag{2.133}$$

and

$$\Lambda_I^{(2)} = \frac{1}{2} \left[\begin{pmatrix} \cosh\phi & -\sinh\phi & 0 & 0 \\ -\sinh\phi & \cosh\phi & 0 & 0 \\ 0 & 0 & 1 & 0 \\ 0 & 0 & 0 & 1 \end{pmatrix} - \begin{pmatrix} \cosh\phi & \sinh\phi & 0 & 0 \\ \sinh\phi & \cosh\phi & 0 & 0 \\ 0 & 0 & 1 & 0 \\ 0 & 0 & 0 & 1 \end{pmatrix} \right]$$

$$= \frac{1}{2} \begin{pmatrix} 0 & 2\sinh\phi & 0 & 0 \\ 2\sinh\phi & 0 & 0 & 0 \\ 0 & 0 & 0 & 0 \\ 0 & 0 & 0 & 0 \end{pmatrix}$$

$$= \sinh\phi \begin{pmatrix} 0 & -1 & 0 & 0 \\ -1 & 0 & 0 & 0 \\ 0 & 0 & 0 & 0 \\ 0 & 0 & 0 & 1 \end{pmatrix} \tag{2.134}$$

With all this, the Frenet-Serret matrix is given by

$$\Theta = \frac{1}{l} \left(\Lambda_R^{(2)} \right)^{-1} \Lambda_I^{(2)}$$

$$= \frac{1}{l} \begin{pmatrix} \frac{1}{\cosh\phi} & 0 & 0 & 0 \\ 0 & \frac{1}{\cosh\phi} & 0 & 0 \\ 0 & 0 & 1 & 0 \\ 0 & 0 & 0 & 1 \end{pmatrix} \begin{pmatrix} 0 & -\sinh\phi & 0 & 0 \\ -\sinh\phi & 0 & 0 & 0 \\ 0 & 0 & 1 & 0 \\ 0 & 0 & 0 & 1 \end{pmatrix}$$

$$= \begin{pmatrix} 0 & -\frac{1}{l}\tanh\phi & 0 & 0 \\ -\frac{1}{l}\tanh\phi & 0 & 0 & 0 \\ 0 & 0 & 1 & 0 \\ 0 & 0 & 0 & 1 \end{pmatrix}. \tag{2.135}$$

Because $-1 \leq \tanh\phi \leq 1$, we have

$$-\frac{1}{l} \leq \Theta^1_0 = -\frac{1}{l}\tanh\phi \leq \frac{1}{l}. \tag{2.136}$$

The Θ^1_0 describe acceleration in the x-direction and (2.136) implies that this acceleration is limited from above by $\frac{1}{l}$.

The Frenet-Serret tensor [25] appears in the Lorentz transformation of Exercise 2.5: Applying $\Lambda^{(2)}$ to an orthonormal frame, given by the vectors $\mathbf{e}_\mu$. The result is

$$\Lambda^{(2)}\mathbf{e}_\mu = \left(\Lambda^{(2)}_R + I\Lambda^{(2)}_I\right)\mathbf{e}_\mu$$

$$= \Lambda^{(2)}_R \left(1 + lI\Theta\right)\mathbf{e}_\mu, \tag{2.137}$$

with the Frenet-Serret tensor

$$\Theta = \frac{1}{l}\left(\Lambda^{(2)}_R\right)^{-1}\Lambda^{(2)}_I. \tag{2.138}$$

As shown above and in [25], this tensor is related to the acceleration of a system.

This procedure is applied to the construction of a real metric for $l = 0$. First, one has to get the same functional form for the metrics $g^\pm_{\mu\nu}$. The same function for $g^\pm_{\mu\nu}$ is constructed as follows: As will be seen, the $g^\pm_{\mu\nu}(X_\pm)$ differ by one term which includes in $g^-_{\mu\nu}$ a function $\Omega_-(X_-, B)$, where B may be one parameter or a set of parameters. Let us write as an example the structure for the pc-Schwarzschild solution (see Chap. 3):

$$g^+_{00} = 1 - \frac{\mathcal{M}_+}{R_+}$$

$$g^-_{00} = 1 - \frac{\mathcal{M}_-}{R_-} + \frac{\Omega_-(R_-, B)}{R_-} \tag{2.139}$$

($\mathcal{M}_\pm$ are integration parameters and identified *both* with m (half the Schwarzschild radius), in order to reproduce for large, but not so large, radial distances the standard Schwarzschild solution) and let us suppose that this function can be written as

$$\Omega_- = \sum_{n=2}^{\infty} \frac{B^-_n}{R^n_-}. \tag{2.140}$$

The restriction to $n \geq 2$ is due to the following reason: In the *Parametrized Post Newtonian Formalism* (PPN) (see for details [35]), a new radial variable $\bar{r}$, called isotropic distance (i.e., the preferred system of coordinates by astronomers are the isotropic coordinates [35]) is defined by

$$r = \bar{r} \left(1 + \frac{m}{2\bar{r}}\right)^2 . \tag{2.141}$$

Considering as a particular example the metric component g_{00} in the standard Schwarzschild solution, it transforms to

$$
\begin{aligned}
\left(1 - \frac{2m}{r}\right) &= 1 - \frac{2m}{\bar{r}\left(1 + \frac{m}{2\bar{r}}\right)^2} \\
&= \frac{\bar{r}\left(1 + \frac{m}{2\bar{r}}\right)^2 - 2m}{\bar{r}\left(1 + \frac{m}{2\bar{r}}\right)^2} \\
&= \frac{\left[1 + \frac{m}{\bar{r}} + \left(\frac{m}{2\bar{r}}\right)^2 - \frac{2m}{\bar{r}}\right]}{\left(1 + \frac{m}{2\bar{r}}\right)^2} \\
&= \frac{\left[1 - \frac{m}{\bar{r}} + \left(\frac{m}{2\bar{r}}\right)^2\right]}{\left(1 + \frac{m}{2\bar{r}}\right)} \\
&= \frac{\left(1 - \frac{m}{2\bar{r}}\right)^2}{\left(1 + \frac{m}{2\bar{r}}\right)^2} .
\end{aligned}
\tag{2.142}
$$

This is expanded in powers of $\frac{m}{2\bar{r}}$. The $\bar{r}$ is for large radial distances proportional to r and thus very large compared to m. For example, for the sun $2m$ is about 3 km, while r is several million kilometers. Therefore, in solar system experiments it suffices to assume the expansion parameter to be very small.

Expanding (2.142) up to the second power yields

$$
\begin{aligned}
\left(1 - \left(\frac{m}{\bar{r}}\right) + \left(\frac{m}{2\bar{r}}\right)^2\right)&\left(1 - 2\left(\frac{m}{2\bar{r}}\right) + 3\left(\frac{m}{2\bar{r}}\right)^2\right) \\
&= 1 - 2\frac{m}{\bar{r}} + 2\left(\frac{m}{\bar{r}}\right)^2 + \cdots .
\end{aligned}
\tag{2.143}
$$

This is the result for the Schwarzschild solution in standard GR. Astronomers change (2.143) into

$$1 - \alpha\frac{m}{\bar{r}} + \beta\left(\frac{m}{\bar{r}}\right)^2 + \cdots , \tag{2.144}$$

introducing the parameters α and β, to be adjusted in solar system experiments. The observations, performed up to now [35, 38]) in the solar system, are consistent with

$\alpha = 2$ and $\beta = 2$. For example, the error for β is smaller than 3×10^{-3}, obtained through the perihelion shift of Mercury.

When in g_{00} a correction term proportional to $\frac{1}{r^2}$ is introduced, due to (2.141) the leading term of that correction in the isotropic radial distance is also proportional to $\frac{1}{\bar{r}^2}$. This is excluded by measurements [38] which showed that the $1/\bar{r}^2$ corrections are extremely well reproduced by Einstein's theory, leaving little space for a contribution of the type $B/\bar{r}^2$ with a large B. Taking into account that no $1/\bar{r}^2$ corrections are measured, a correction with a sufficient large parameter in order to erase the event horizon can only start with $\frac{1}{\bar{r}^n}$, with $n \geq 3$.

In what follows, we will introduce a notation which allows us to write the $\sigma_\pm$ components of the metric in the same functional form, adding terms which are zero because the new parameters added are zero. The final form has a much nicer symmetry between both zero-divisor components and allows to define the projection:

Due to the just exposed argument, we make the following ansatz for the metric: To the $g_{\mu\nu}^+(X_+)$ in (2.139) we add *the same* function as in (2.140), with the minus sign changed to the plus sign. Defining the pc-parameters

$$B_n = B_n^+ \sigma_+ + B_n^- \sigma_- = B_{n_-} \sigma_-, \tag{2.145}$$

allows us to write the pc-parameter B_n in a symmetric form, by adding a parameter B_{n_+} in the σ_+ component. Of course the B_{n_+} parameter has to be set later on to zero, in order to obtain our previous result. To the term in the σ_- component, one is allowed to add the same form of the term in the σ_+ component, namely

$$\Omega_+ = \sum_{n=2}^{\infty} \frac{B_n^+}{R_+^n} \tag{2.146}$$

to $g_{\mu\nu}^+$, thus yielding the same functional form as $g_{\mu\nu}^-$. When the mapping to the real metric is performed, we proceed as in the case of the Lorentz transformations, setting $X_\pm^\mu \to x^\mu$ and all parameters $B_n \to \mathrm{Re}(B_n) = \frac{1}{2}\left(B_n^+ + B_n^-\right) = \frac{1}{2}B_{n_-}$ (because B_{n_+} is zero).

In order to relate the theory to already existing ones, we estimate the dominant contributions in the length element square, i.e., for the moment we still use $l > 0$.

Once the pc-metric $g_{\mu\nu}(X, \mathscr{A})$ is obtained, the above proposed rule of the mapping results into

$$g_{\mu\nu}(X, \mathscr{A}) \to g_{\mu\nu}(x, \mathscr{A}_R), \tag{2.147}$$

where the index R refers to the real component. The justification for (2.147) is as follows: The pc-variables also depend on lu^μ. Because the four velocity is limited by c, this expression is very small and one can approximate the metric by expanding it in terms of powers in lu^μ, i.e.,

$$g_{\mu\nu}(X, \mathscr{A}) \approx g_{\mu\nu}(x, \mathscr{A}) + f_{\mu\lambda}(x, \mathscr{A})lu^\lambda. \tag{2.148}$$

Because $u^\lambda \leq c$ and the minimal length l is probably of the order of the Planck length, it implies that the contribution of u^μ and l to the metric can be safely neglected. Finally, we substitute the parameters by their real component, as was discussed further above when the pc-Lorentz transformation was investigated and mapped to the standard Lorentz transformation. There we defined the mapping to the four-dimensional subspace, where the contributions of the minimal length are neglected.

The argument just exposed can not be applied to *differences* in velocity du^μ, representing the acceleration. In the vicinity of a very large mass the acceleration becomes very large. When the acceleration approaches its maximal possible value, then such terms including du^μ are getting important. However, as long as the acceleration is small, compared to the maximal one, we also can skip the l-dependent terms.

This is exploited in the length element squared (2.78), which now becomes

$$
\begin{aligned}
d\omega^2 &= g_{\mu\nu}(X, \mathscr{A}) DX^\mu DX^\nu \\
&= g_{\mu\nu}(X, \mathscr{A}) \left(dx^\mu dx^\nu + l^2 du^\mu du^\nu\right) + g_{\mu\nu}(X, \mathscr{A}) 2 l I dx^\mu du^\nu \\
&\rightarrow g_{\mu\nu}(x, \mathscr{A}_R) \left(dx^\mu dx^\nu + l^2 du^\mu du^\nu\right) + g_{\mu\nu}(x, \mathscr{A}_R) 2 l I dx^\mu du^\nu \\
&\rightarrow g_{\mu\nu}(x, \mathscr{A}_R) \left(dx^\mu dx^\nu + l^2 du^\mu du^\nu\right).
\end{aligned}
\tag{2.149}
$$

In the first step the pc-variables were substituted and in a second step (third line) the metric components were approximated in the limit of $l = 0$. In the fourth line the mixed term $dx^\mu du^\nu$ is skipped, because we require that the length element of the motion of a physical particle has to be real. This is only the case when the pseudo-imaginary component in (2.149) is set to zero. *This condition is nothing else as the dispersion relation, which results as a necessary by-product of the pc-formulation.* Most authors add this condition by hand.

Note, that (2.149) is the length element squared as used in [6, 12–18]. Thus, all the former mentioned theories with a minimal length parameter l are contained in pc-GR as a special limit!

The procedure just outlined, did lead to the metrics as discussed in the pc-Schwarzschild, pc-Reissner-Nortström and pc-Kerr solution [32, 33] (Chap. 3) and also of the pc-Robertson-Walker Model of the universe [34] (Chap. 4).

At the end of this chapter we resume the main results:

- The main features of the pc-GR is that it includes a *minimal length* and that due to the modified variational principle there is always a contribution to the equation of motion, which we interpret as a dark energy. This contribution has to be different from zero. A justification of the modified variational principle was given through the implementation of a real valued length element, using the Lagrange multiplyer formalism.
- It was shown that for a very small minimal length but allowing large accelerations the theory reduces to the ones discussed by Caianiello [6] and more subsequent contributions by others. Also the length element is equal to the one in hermitian gravities. Thus, all these theories are contained in pc-GR, which thus can be considered as a covering theory.

- The pc-GR treats coordinates and velocities on the same level, thus giving the possibility to satisfy Born's reciprocity condition [4, 5].
- This shows that pc-GR is more general and the investigation of the influence of the minimal length is necessary in order to exploit the most general form of pc-GR.
- In practical applications, however, the minimal length and the acceleration are too small, as that one could observe an effect. Only when the metric becomes nearly singular, the minimal length might play an observable role. The part of pc-GR which remains is the contribution of the dark energy, which has to be there due to the modified variational principle. About these applications will treat the next chapters.

References

1. M.A. Abramowicz, W. Klu, No observational proof of the black-hole event horizon. A&A **396**, L31 (2002)
2. A. Einstein, A generalization of the relativistic theory of gravitation. Ann. Math. Second Ser. **46**, 578 (1945)
3. A. Einstein, A generalized theory of gravitation. Rev. Mod. Phys. **20**, 35 (1948)
4. M. Born, A suggestion for unifying quantum theory and relativity. Proc. Roy. Soc. A **165**, 291 (1938)
5. M. Born, Reprocity theory of elementary particles. Rev. Mod. Phys. **21**, 463 (1949)
6. E.R. Caianiello, Is there a maximal acceleration? Nuovo Cim. Lett. **32**, 65 (1981)
7. S.W. Hawking, G.F.R. Ellis, *The Large Scale Structure of Space-time* (Cambridge University Press, 1973)
8. R.M. Wald, *General Relativity* (University of Chicago Press, 1994)
9. G. Kunstatter, R. Yates, The geometrical structure of a complexified theory of gravitation. J. Phys. A **14**, 847 (1981)
10. V. Bozza, A. Feoli, G. Lambiase, G. Papini, G. Scarpetta, Maximal acceleration effects in Kerr space. Phys. Lett. A **283**, 847 (2001)
11. S. Capozziello, A. Feoli, G. Lambiase, G. Papini, G. Scarpetta, Massive scalar particles in a modified Schwarzschild geometry. Phys. Lett. A **268**, 247 (2000)
12. R.G. Beil, Electrodynamics from a metric. Int. J. Theor. Phys. **26**, 189 (1987)
13. R.G. Beil, New class of finsler metrics. Int. J. Theor. Phys. **28**, 659 (1989)
14. R.G. Beil, Finsler gauge transformations and general relativity. Int. J. Theor. Phys. **31**, 1025 (1992)
15. R.G. Beil, Finsler geometry and relativistic field theory. Found. Phys. **33**, 110 (2003)
16. H.E. Brandt, Maximal proper acceleration and the structure of spacetime. Found. Phys. Lett. **2**, 39 (1989)
17. H.E. Brandt, Structure of spacetime tangent bundles. Found. Phys. Lett. **4**, 523 (1989)
18. H.E. Brandt, Complex spacetime tangent bundle. Found. Phys. Lett. **6**, 245 (1993)
19. S.G. Low, Canonical relativistic quantum mechanics: representations of the unitary semidirect Heisenberg group, $U(1, 3) \times_S H(1, 3)$. J. Math. Phys. **38**, 2197 (1997)
20. C.L.M. Mantz, T. Prokopec, *Hermitian Gravity and Cosmology* (2008). arXiv:0804.0213
21. C.L.M. Mantz, T. Prokopec, Resolving curvature singularities in holomorphic gravity. Found. Phys. **41**, 1597 (2011)
22. W. Moffat, A new theory of gravitation. Phys. Rev. D **19**, 3554 (1979)
23. P.F. Kelly, R.B. Mann, Ghost properties of algebraically extended theories of gravitation. Class. Quantum Gravity **3**, 705 (1986)
24. W. Greiner, J. Reinhardt, *Field quantization* (Springer, Heidelberg, 1993)

25. R. Adler, M. Bazin, M. Schiffer, *Introduction to General Relativity*, 2nd edn. (McGraw Hill, New York, 1975)
26. S. Carroll, *Spacetime and Geometry. An Introduction to General Relativity* (Addison-Wesley, San Francisco, 2004)
27. P.O. Hess, W. Greiner, Pseudo-complex field theory. Int. J. Mod. Phys. E **16**, 1643 (2007)
28. C. Barceló, S. Liberati, S. Sonego, M. Visser, Fate of gravitational collapse in semiclassical gravity. Phys. Rev. D **77**, 044032 (2008)
29. M. Visser, Gravitational vacuum polarization I: energy conditions in the Hartle-Hawking vacuum. Phys. Rev. D **54**, 5103 (1996)
30. M. Visser, Gravitational vacuum polarization II: energy conditions in the Bouleware vacuum. Phys. Rev. D **54**, 5116 (1996)
31. M. Visser, Gravitational vacuum polarization IV: energy conditions in the Unruh vacuum. Phys. Rev. D **56**, 936 (1997)
32. G. Caspar, T. Schönenbach, P.O. Hess, M. Schäfer, W. Greiner, Pseudo-complex general relativity: Schwarzschild, Reissner-Nordstrøm and Kerr solutions. Int. J. Mod. Phys E. **21**, 1250015 (2012)
33. P.O. Hess, W. Greiner, Pseudo-complex general relativity. Int. J. Mod. Phys. E **18**, 51 (2009)
34. P.O. Hess, L. Maghlaoui, W. Greiner, The Robertson-Walker metric in a pseudo-complex general relativity. Int. J. Mod. Phys. E **19**, 1315 (2010)
35. C.W. Misner, K.S. Thorne, J.A. Wheeler, *Gravitation* (Freeman & Co., San Francisco, 1973)
36. F.P. Schuller. *Dirac-Born-Infeld Kinematics, Maximal Acceleration and Almost Product Manifolds*. Ph.D. thesis, University of Cambridge, 2003
37. F.P. Schuller, M.N.R. Wohlfarth, T.W. Grimm, Pauli Villars regularization and Born Infeld kinematics. Class. Quantum Gravity **20**, 4269 (2003)
38. C.M. Will, The confrontation between general relativity and experiment. Living Rev. Relativ. **9**, 3 (2006)

Chapter 3
Solutions for Central Masses: pc-Schwarzschild, pc-Kerr and pc-Reissner-Nordström

In this chapter the solutions for systems with spherical or axial symmetries are presented within the pseudo-complex (pc) General Relativity (pc-GR). First a non-rotating central mass is considered, which is the pc-Schwarzschild solution. The steps for solving the Einstein equations are in line to those presented in [1], with some new ingredients. In the second part, the solution for a rotating central mass will be discussed, which is the pc-Kerr solution. There, one has to take a different path as presented in [1] due to the complexities involved in its derivation and the additional presence of an energy-momentum tensor representing the dark energy. Finally, we discuss the pc-Reissner-Nordström solution, which describes a charged central mass. The main source of reference for all cases is [2].

In order to follow this chapter, we recommend to study the first two chapters.

3.1 Modified Variational Procedure in an 8-dimensional Space

Working in the 8-dimensional pseudo-complex space, one has to consider a change in the variational procedure, which is also applicable when the minimal length l, emerging in this theory, is set to zero.

We apply the modified variational principle as introduced in Chap. 2, i.e.,

$$\delta S = \delta \int dX^4 \sqrt{-g}\mathscr{R},\qquad(3.1)$$

where g is the determinant of the metric, R the Riemann scalar and both are pseudo-complex functions. The variation of the pc-action is proportional to an element in the zero-divisor (see Chap. 2), or

$$\delta S = \lambda \sigma_-.\qquad(3.2)$$

© Springer International Publishing Switzerland 2016
P.O. Hess et al., *Pseudo-Complex General Relativity*,
FIAS Interdisciplinary Science Series, DOI 10.1007/978-3-319-25061-8_3

The λ-function gives us a great liberty, which will be used to eliminate the event horizon of a formerly black hole. In fact it must be constructed such that our general principle of a singularity-free theory holds. This modified variational principle also leads to the appearance of the dark energy tensor which is *non-zero*, otherwise one arrives at two independent theories, as stated above.

After variation one obtains

$$\mathscr{R}_{\mu\nu} - \frac{1}{2} g_{\mu\nu} \mathscr{R} = \frac{8\pi\kappa}{c^2} \left(T^{\Lambda} \right)_{\mu\nu} \sigma_-, \tag{3.3}$$

where on the right hand side the factor $\frac{8\pi\kappa}{c^2}$ is intentionally extracted in order to identify the contribution from the modified variational principle as an additional energy-momentum tensor. The superscript Λ indicates that we will identify this contribution with a dark energy.

For convenience, we assume that the functional dependence of the metric $g^+_{\mu\nu}(X_+)$ in terms of X^λ_+ is the same as $g^-_{\mu\nu}(X_-)$ in terms of X^λ_-. In case when $g^-_{\mu\nu}$ differs only by one term with respect to $g^+_{\mu\nu}$, in particular by a function with the structure

$$F_{\mu\nu}(R_-) = \sum_n \frac{\left(B^-_n \right)_{\mu\nu}}{R^n_-}, \tag{3.4}$$

where we assumed that this function is expressible in terms of a Laurent series, we can add to $g^+_{\mu\nu}$ (for the case of the central mass problem, which we are mainly interested in) the function

$$F_{\mu\nu}(R_+) = \sum_n \frac{\left(B^+_n \right)_{\mu\nu}}{R^n_+}, \tag{3.5}$$

and define

$$(B_n)_{\mu\nu} = \left(B^+_n \right)_{\mu\nu} \sigma_+ + \left(B^-_n \right)_{\mu\nu} \sigma_- \tag{3.6}$$

for the expansion coefficients, with $\left(B^+_n \right)_{\mu\nu} = 0$. In this manner, a zero is added to $g^+_{\mu\nu}$. Due to this trick, the functional dependence is now formally equal in both zero-divisor components.

For $l = 0$, or assuming that the contributions of the minimal length are very small, we expect

$$X^\mu_\pm \to x^\mu$$
$$g^\pm_{\mu\nu}(X_\pm) \to g_{\mu\nu}(x)$$
$$\mathscr{R}^\pm_{\mu\nu} \to \mathscr{R}_{\mu\nu}$$
$$\mathscr{R}^\pm \to \mathscr{R}. \tag{3.7}$$

For the pseudo-complex parameters $(B_n)_{\mu\nu} = \left(\left(B_n^+ \right)_{\mu\nu} \sigma_+ + \left(B_n^- \right)_{\mu\nu} \sigma_- \right)$ we have $\left(B_n^+ \right)_{\mu\nu} = 0$ and, thus, $(B_n)_{\mu\nu} = \left(B_n^- \right)_{\mu\nu} \sigma_-$. The pseudo-real part of $\sigma_- = \frac{1}{2}(1-I)$ is $\frac{1}{2}$. Therefore, the mapping of the pc-parameters yields

$$(B_n)_{\mu\nu} \rightarrow \frac{1}{2} \left(B_n^- \right)_{\mu\nu}. \tag{3.8}$$

On the right hand side of (3.3) there is an additional contribution of an energy-momentum tensor, which is a consequence of the pc-description. Since we require a vanishing singularity, the functional dependence of $\left(T^\Lambda \right)_{\mu\nu}$ on x^λ has to be constructed such, that the singularity indeed vanishes.

3.1.1 Isotropic and Anisotropic Fluids

Two different examples for the dark energy will be discussed, namely a fluid model which is either isotropic or anisotropic. As will be seen further below, for the fluid outside the star *an anisotropic fluid has to be assumed*, otherwise the only solution is a constant density. For an isotropic fluid the standard expression holds, namely

$$T_{\mu\nu} = \varepsilon u_\mu u_\nu + \frac{p}{c^2} \left(u_\mu u_\nu + g_{\mu\nu} \right)$$
$$= \left(\varepsilon + \frac{p}{c^2} \right) u_\mu u_\nu + \frac{p}{c^2} \delta_{\mu\nu}. \tag{3.9}$$

The 4-velocity satisfies the condition $g_{\mu\nu} u^\mu u^\nu = -1$, with the choice of $u^0 = \frac{1}{\sqrt{|g_{00}|}}$ and $u^\mu = 0$, for $\mu \neq 0$ (see Exercise 6.1 of Chap. 6).

The energy-momentum tensor acquires the form (see Exercise 6.1 of Chap. 6)

$$T_\mu^\nu = \begin{pmatrix} -\varepsilon & 0 & 0 & 0 \\ 0 & \frac{p}{c^2} & 0 & 0 \\ 0 & 0 & \frac{p}{c^2} & 0 \\ 0 & 0 & 0 & \frac{p}{c^2} \end{pmatrix}, \tag{3.10}$$

with $T_\mu^\nu = g^{\nu\lambda} T_{\mu\lambda}$.

For an anisotropic fluid, the energy momentum tensor is

$$T_{\mu\nu} = (\varepsilon + \frac{p_\vartheta}{c^2}) u_\mu u_\nu + \frac{p_\vartheta}{c^2} g_{\mu\nu} + \frac{1}{c^2}(p_r - p_\vartheta) k_\mu k_\nu, \tag{3.11}$$

with k^μ being a space-like vector satisfying $k^\mu u_\mu = 0$ and $k^\mu l_\mu = 0$, with l_μ a null-vector. The ansatz can be explained as follows: The u^μ is a time-like vector. In order to have an anisotropy, one has to add a term similar to $u_\mu u_\nu$, but with a space-like vector k_μ, which is by construction orthogonal to u_μ. The p_r is the radial component and p_ϑ the tangential component.

The factors in front of the terms in (3.11) are obtained when one considers a particular coordinate system, in which the 4-velocity has only a component in time and the space-like vector has only a radial component k^1. In this special frame we have $g_{\mu\nu}u^\mu u^\nu = -1 = g_{00}\left(u^0\right)^2$ and $g_{\mu\nu}k^\mu k^\nu = 1 = g_{00}\left(v^1\right)^2$. Thus, $u^0 = 1/\sqrt{|g_{00}|}$, $u^k = 0$ ($k = 1, 2, 3$) and $k^k = 1/\sqrt{g_{11}}$, $k^\mu = 0$ with $\mu \neq k$ ($k = 2, 3$), having used $g_{\mu\nu}k^\mu k^\nu = +1$, this energy-momentum tensor acquires the form

$$T^\nu_\mu = \begin{pmatrix} -\varepsilon & 0 & 0 & 0 \\ 0 & \frac{p_r}{c^2} & 0 & 0 \\ 0 & 0 & \frac{p_\vartheta}{c^2} & 0 \\ 0 & 0 & 0 & \frac{p_\vartheta}{c^2} \end{pmatrix}. \tag{3.12}$$

It is easy to verify that (3.12) reduces to (3.10) when $p_\vartheta = p_r$.

In the following applications we always will assume the pressure and density to be real.

3.2 The pc-Schwarzschild Solution

This metric is time-independent with a radial symmetry. Therefore, its general form is

$$D\omega^2 = -e^\nu c^2 Dt^2 + e^\lambda DR^2 + R^2\left(\sin^2\vartheta D\varphi^2 + D\vartheta^2\right), \tag{3.13}$$

where $\nu(R)$ and $\lambda(R)$ are free function, depending only on the radial coordinate. These functions have to be derived from the Einstein equations. In Einstein's equation occurs the curvature tensor, which in turn depends on the Christoffel symbols, so we have to start with their calculation.

This is achieved most easily via the use of the geodesic equation, for the motion of a test particle,

$$\ddot{X}^\mu + \left\{ \begin{matrix} \mu \\ \nu\lambda \end{matrix} \right\} \dot{X}^\nu \dot{X}^\lambda = 0, \tag{3.14}$$

where the coefficients of the equation identify the Christoffel symbols. Note, that the metric is the solution of the Einstein equations (3.3). Once obtained, the question to find the geodesic is a standard variational procedure.

The action to vary is

$$\delta \int \sqrt{-D\omega^2} = \delta \int \sqrt{\left(-\frac{D\omega}{D\omega}\right)} d\omega = 0. \tag{3.15}$$

where ω is a pc-curve parameter.

A straightforward calculation yields (see Exercise 3.1)

$$\ddot{X}^0 + \nu' \dot{R} \dot{X}^0 = 0,$$

$$\ddot{R} + \frac{1}{2}\lambda' \dot{R}^2 + \frac{1}{2}\nu' e^{\nu-\lambda}(\dot{X}^0)^2 - e^{-\lambda} R \dot{\vartheta}^2$$

$$-R\sin^2\vartheta \dot{\varphi}^2 e^{-\lambda} = 0,$$

$$\ddot{\vartheta} + \frac{2}{R}\dot{\vartheta}\dot{R} - \sin\vartheta\cos\vartheta \dot{\varphi}^2 = 0,$$

$$\ddot{\varphi} + 2\cot\vartheta \dot{\varphi}\dot{\vartheta} + \frac{2}{R}\dot{R}\dot{\varphi} = 0. \qquad (3.16)$$

A prime indicates the derivative with respect to R, e.g., $\nu' = \frac{D\nu}{DR}$.

Exercise 3.1 (Proof of (3.16))

Problem. The Lagrange function to vary is

$$L = -\left(\frac{D\omega}{D\omega}\right)^2$$

$$= e^\nu \left(\dot{X}^0\right)^2 - e^\lambda \dot{R}^2 - R^2 \left(\sin^2\vartheta \dot{\varphi}^2 + \dot{\vartheta}^2\right), \qquad (3.17)$$

with $X^0 = ct$. A dot denotes the derivative with respect to s. The variational problem is equivalent to (3.15).

The derivatives of the Lagrangian with respect to the velocities and coordinates are

$$\frac{DL}{D\dot{X}^0} = 2e^\nu \dot{X}^0, \quad \frac{DL}{DX^0} = 0,$$

$$\frac{DL}{D\dot{R}} = -2e^\lambda \dot{R}, \quad \frac{DL}{DR} = \nu' e^\nu \left(\dot{X}^0\right)^2 - \lambda e^\lambda \dot{R}^2 - 2R\dot{R}\left(\sin^2\vartheta \dot{\varphi}^2 + \dot{\vartheta}^2\right),$$

$$\frac{DL}{D\dot{\varphi}} = -2R^2\sin^2\vartheta \dot{\varphi}, \quad \frac{DL}{D\varphi} = 0,$$

$$\frac{DL}{D\dot{\vartheta}} = -2R^2\dot{\vartheta}, \quad \frac{DL}{D\vartheta} = -2R^2\sin\vartheta\cos\vartheta \dot{\varphi}^2. \qquad (3.18)$$

The prime indicates the derivative with respect to R.

Consider as an example the equation of the X^0 component. Using (3.18) the Euler-Lagrange equation reads

$$\frac{D}{Ds}\frac{DL}{D\dot{X}^0} - \frac{DL}{DX^0} = \frac{D\left(2e^{\nu}\dot{X}^0\right)}{ds}$$
$$= 2\nu'e^{\nu}\dot{R}\dot{X}^0 + 2e^{\nu}\ddot{X}^0 = 0. \qquad (3.19)$$

Dividing by $2e^{\nu}$ gives the first equation in (3.16).

The derivative of ν or λ with respect to ω, i.e. $\dot{\nu}$ and $\dot{\lambda}$, can be written as

$$\dot{\nu} = \frac{D\nu}{DR}\dot{R} = \nu'\dot{R} \qquad (3.20)$$

and similar for $\dot{\lambda}$.

In a similar fashion all the other equations in (3.16) are derived.

Comparing (3.16) with (3.14), yields for the Christoffel symbols;

$$\begin{Bmatrix} 0 \\ 10 \end{Bmatrix} = \frac{1}{2}\nu' = \begin{Bmatrix} 0 \\ 01 \end{Bmatrix},$$

$$\begin{Bmatrix} 1 \\ 00 \end{Bmatrix} = \frac{1}{2}\nu'e^{\nu-\lambda},$$

$$\begin{Bmatrix} 1 \\ 11 \end{Bmatrix} = \frac{1}{2}\lambda',$$

$$\begin{Bmatrix} 1 \\ 22 \end{Bmatrix} = -Re^{-\lambda},$$

$$\begin{Bmatrix} 1 \\ 33 \end{Bmatrix} = -R\sin^2\vartheta\, e^{-\lambda},$$

$$\begin{Bmatrix} 2 \\ 21 \end{Bmatrix} = \frac{1}{R} = \begin{Bmatrix} 2 \\ 12 \end{Bmatrix},$$

$$\begin{Bmatrix} 2 \\ 33 \end{Bmatrix} = -\sin\vartheta\, \cos\vartheta,$$

$$\begin{Bmatrix} 3 \\ 23 \end{Bmatrix} = \cot\vartheta = \begin{Bmatrix} 3 \\ 32 \end{Bmatrix},$$

$$\begin{Bmatrix} 3 \\ 13 \end{Bmatrix} = \frac{1}{R} = \begin{Bmatrix} 3 \\ 31 \end{Bmatrix}, \qquad (3.21)$$

which can be readily copied from [1], with the difference that now the variables R, ϑ and φ are pseudo-complex. Though, the ϑ and the φ are in general pseudo-complex, when we consider in both components of the zero divisor an identical spherical problem, these angles can be set to be pseudo-real.

The Christoffel symbols are used to derive the Riemann curvature and Ricci tensor, which appear in the modified Einstein equations

$$\mathscr{R}_\mu{}^\nu - \frac{1}{2} g_\mu{}^\nu \mathscr{R} = \frac{8\pi\kappa}{c^2} \left(T^\Lambda\right)_\mu{}^\nu \sigma_- \equiv \Xi_\mu{}^\nu \sigma_-, \tag{3.22}$$

where we introduced a shorthand notation on the right hand side.

When additional matter distribution is present, then one has to add on the right hand side of the Einstein equations

$$\frac{8\pi\kappa}{c^2} \left(T^{\mathrm{mat}}\right)_\mu{}^\nu = \frac{8\pi\kappa}{c^2} \left(T^{\mathrm{mat}}\right)_\mu{}^\nu (\sigma_+ + \sigma_-), \tag{3.23}$$

where $(\sigma_+ + \sigma_-) = 1$ has been used. For this case, in each zero-divisor component there appears the same contribution of the matter, indicated by "mat". More on this will be presented in Chap. 6, dedicated to the description of neutron stars, where matter is present in its interior.

The solution of the σ_+ part is obtained as in the standard GR. Both metric components are given by

$$(g_{\mu\nu}^\pm) = \begin{pmatrix} -\left(1 - \frac{2\mathscr{M}_\pm}{R_\pm} + \frac{\Omega_\pm}{R_\pm}\right) e^{f_\pm} & 0 & 0 & 0 \\ 0 & \left(1 - \frac{2\mathscr{M}_\pm}{R_\pm} + \frac{\Omega_\pm}{R_\pm}\right)^{-1} & 0 & 0 \\ 0 & 0 & R_\pm^2 & 0 \\ 0 & 0 & 0 & R_\pm^2 \sin^2\theta \end{pmatrix}. \tag{3.24}$$

The functions $f_\pm$ arise in general when $(\nu' + \lambda')$ is not zero. An explicit example can be found at the end of this chapter, or in Chap. 6. The relations to the formerly defined functions $\nu(r)$ and $\lambda(r)$ are

$$e^{\nu_\pm(R_\pm)} = \left(1 - \frac{2\mathscr{M}_\pm}{R_\pm} + \frac{\Omega_\pm}{R_\pm}\right) e^{f_\pm}$$

$$e^{\lambda_\pm(R_\pm)} = \left(1 - \frac{2\mathscr{M}_\pm}{R_\pm} + \frac{\Omega_\pm}{R_\pm}\right)^{-1}. \tag{3.25}$$

The $\Omega_\pm$ functions are supposed to be expressible in an expansion in $1/R_\pm$. The expansion coefficients $(B_n)_+$ are set to zero, as explained further above. The B_{n+} serve only to remember that the same functional form for both metric components is formally obtained.

For the σ_- component, the inclusion of the additional dark energy-momentum tensor has to be considered, which will be done now.

For the solution, one has to keep in mind that on the right hand side of the Einstein equations there is a zero divisor contribution, proportional to σ_-. The Ricci tensor appears on the left hand side of the Einstein equations, where the right hand side is proportional to σ_-. Each $\mathscr{R}_{\mu\nu}$ gives a contribution proportional to σ_- to the right hand side of the Einstein equations. We separate these contributions and associate to

each component of the Ricci tensor the following contribution to the right hand side
of the Einstein equations:

$$\mathscr{R}_{00} = \frac{1}{2} e^{\nu - \lambda} \xi_0 \sigma_-$$

$$\mathscr{R}_{11} = -\frac{1}{2} \xi_1 \sigma_-$$

$$\mathscr{R}_{22} = -\xi_2 \sigma_-$$

$$\mathscr{R}_{33} = -\xi_3 \sigma_- = -\xi_2 \sin^2 \vartheta \, \sigma_-. \tag{3.26}$$

where the functions ξ_μ have been introduced and the relation $\mathscr{R}_{33} = \mathscr{R}_{22} \sin^2 \vartheta$ was
used.

The curvature scalar $\mathscr{R}$ on the left hand side of the Einstein equations is given by

$$
\begin{aligned}
\mathscr{R} &= \mathscr{R}_0^0 + \mathscr{R}_1^{\,1} + \mathscr{R}_2^{\,2} + \mathscr{R}_3^{\,3} \\
&= g^{00} \mathscr{R}_{00} + g^{11} \mathscr{R}_{11} + g^{22} \mathscr{R}_{22} + g^{33} \mathscr{R}_{33} \\
&= -e^{-\nu} \mathscr{R}_{00} + e^{-\lambda} \mathscr{R}_{11} + \frac{\mathscr{R}_{22}}{R^2} + \frac{\mathscr{R}_{33}}{R^2 \sin^2 \vartheta}.
\end{aligned}
\tag{3.27}
$$

For the (00) component, the left hand side of the Einstein equation reads

$$
\begin{aligned}
\mathscr{R}_0^{\,0} - \frac{1}{2} g_0^{\,0} \mathscr{R} &= \frac{1}{2} \left(\mathscr{R}_0^{\,0} - \mathscr{R}_1^{\,1} \mathscr{R}_2^{\,2} \mathscr{R}_3^{\,3} \right) \\
&= -\frac{1}{2} \left(e^{-\nu} \mathscr{R}_{00} + e^{-\lambda} \mathscr{R}_{11} + \frac{1}{R^2} \mathscr{R}_{22} + \frac{1}{R^2 \sin^2 \vartheta} \mathscr{R}_{22} \right). \tag{3.28}
\end{aligned}
$$

Substituting (3.26) into this equations and restricting for a while to the σ_- component, yields

$$-\frac{1}{4} e^{-\lambda_-} \xi_0 + \frac{1}{4} e^{-\lambda_-} \xi_1 + \frac{\xi_2}{R_-^2} = \Xi_0^0. \tag{3.29}$$

Repeating these steps for the other components of the Einstein equations, i.e.
substituting (3.26) into the σ_- component of the Einstein equations, we obtain a
relation between these ξ-functions and the $\Xi_\mu = \Xi_\mu^{\,\mu} = \frac{8\pi\kappa}{c^2} T_\mu^{\,\mu}$:

$$-\frac{1}{4} e^{-\lambda_-} \xi_0 + \frac{1}{4} e^{-\lambda_-} \xi_1 + \frac{\xi_2}{R_-^2} = \Xi_0$$

$$\frac{1}{4} e^{-\lambda_-} \xi_0 - \frac{1}{4} e^{-\lambda_-} \xi_1 + \frac{\xi_2}{R_-^2} = \Xi_1$$

$$\frac{1}{4} e^{-\lambda_-} (\xi_0 + \xi_1) = \Xi_2$$

$$\Xi_3 = \Xi_2. \tag{3.30}$$

Resolving for ξ_μ we obtain

$$\frac{2\xi_2}{R_-^2} = \varXi_0 + \varXi_1$$

$$\frac{1}{2}e^{-\lambda_-}(\xi_0 - \xi_1) = \varXi_1 - \varXi_0$$

$$\frac{1}{4}e^{-\lambda_-}(\xi_0 + \xi_1) = \varXi_2. \tag{3.31}$$

Multiplying the last equation with 2 and adding to it the second equation and subsequently subtracting the second equation from the last one, we obtain

$$e^{-\lambda_-}\xi_0 = 2\varXi_2 + \varXi_1 - \varXi_0$$
$$e^{-\lambda_-}\xi_1 = 2\varXi_2 - \varXi_1 + \varXi_0$$
$$\frac{2\xi_2}{R_-^2} = \varXi_0 + \varXi_1. \tag{3.32}$$

Compared with the Schwarzschild result according to Einstein's theory, additional dark energy terms appear. In general the dark energy will be described by an anisotropic fluid. In the specific case of an isotropic fluid model, both pressures are set equal. The assumption of an anisotropic fluid is not so far fetched and has already been discussed in the literature [3–6].

We use for $\left(T_\mu{}^\nu\right) = g^{\nu\lambda}T_{\mu\lambda} = \mathrm{diag}(-\varepsilon, \frac{p_r}{c^2}, \frac{p_\vartheta}{c^2}, \frac{p_\vartheta}{c^2})$ the expression for an ideal anisotropic fluid/gas [1] and obtain

$$\varXi_0^0 = \varXi_0 = g^{00}\varXi_{00} = -\frac{8\pi\kappa}{c^2}\varepsilon,$$

$$\varXi_1^1 = \varXi_1 = g^{11}\varXi_{11} = \frac{8\pi\kappa}{c^2}\frac{p_r}{c^2},$$

$$\varXi_2^2 = \varXi_2 = \varXi_3^3 = \varXi_3 = g^{33}\varXi_{33} = \frac{8\pi\kappa}{c^2}\frac{p_\vartheta}{c^2}. \tag{3.33}$$

Substituting this into (3.32) gives

$$\frac{2\xi_2}{R_-^2} = \frac{8\pi\kappa}{c^2}\left(-\varepsilon + \frac{p_r}{c^2}\right)$$

$$e^{-\lambda_-}\xi_0 = \frac{8\pi\kappa}{c^2}\left(2\frac{p_\vartheta}{c^2} + \frac{p_r}{c^2} + \varepsilon\right)$$

$$e^{-\lambda_-}\xi_1 = \frac{8\pi\kappa}{c^2}\left(-\varepsilon + 2\frac{p_\vartheta}{c^2} - \frac{p_r}{c^2}\right). \tag{3.34}$$

Now, we return to consider both zero-divisor component again. As shown in Exercise 3.2, we obtain $(R_{\mu\nu} = R_\mu{}^\lambda g_{\lambda\nu})$

$$\mathscr{R}_{00} = \frac{e^{\nu-\lambda}}{2}\left(\nu'' + \frac{\nu'^2}{2} - \frac{\lambda'\nu'}{2} + \frac{2\nu'}{R}\right)$$

$$\mathscr{R}_{11} = -\frac{1}{2}\left(\nu'' + \frac{\nu'^2}{2} - \frac{\lambda'\nu'}{2} - \frac{2\lambda'}{R}\right)$$

$$\mathscr{R}_{22} = -\left(e^{-\lambda}R\right)' + 1 - Re^{-\lambda}\left(\frac{\nu'+\lambda'}{2}\right)$$

$$\mathscr{R}_{33} = -\sin^2\vartheta\left[\left(e^{-\lambda}R\right)' - 1 + Re^{-\lambda}\left(\frac{\nu'+\lambda'}{2}\right)\right]. \tag{3.35}$$

The Ricci scalar is defined by

$$\mathscr{R} = \mathscr{R}_\mu^{\;\mu}. \tag{3.36}$$

Exercise 3.2 (Determination of $\mathscr{R}_{\mu\nu}$)

Problem. Determine the relations of $R_{\mu\nu}$, enlisted in (3.35)

Solution.
   The Ricci tensor is given in terms of the Christoffel symbols as [7]

$$\mathscr{R}_{\mu\nu} = -\left\{\begin{matrix}\beta\\\beta\nu\end{matrix}\right\}_{|\mu} + \left\{\begin{matrix}\beta\\\mu\nu\end{matrix}\right\}_{|\beta}$$
$$-\left\{\begin{matrix}\beta\\\tau\mu\end{matrix}\right\}\left\{\begin{matrix}\tau\\\beta\nu\end{matrix}\right\}$$
$$+\left\{\begin{matrix}\beta\\\tau\beta\end{matrix}\right\}\left\{\begin{matrix}\tau\\\mu\nu\end{matrix}\right\}. \tag{3.37}$$

   We now consider step by step the different components of the Ricci tensor, using (3.37) and that only diagonal components appear.

(1) $\mu = \nu = 0$:
   According to (3.37), he $\mathscr{R}_{00}$ component is given by

$$-\left\{\begin{matrix}\beta\\\beta 0\end{matrix}\right\}_{|0} + \left\{\begin{matrix}\beta\\00\end{matrix}\right\}_{|\beta}$$
$$-\left\{\begin{matrix}\beta\\\tau 0\end{matrix}\right\}\left\{\begin{matrix}\tau\\\beta 0\end{matrix}\right\}$$
$$+\left\{\begin{matrix}\beta\\\tau\beta\end{matrix}\right\}\left\{\begin{matrix}\tau\\00\end{matrix}\right\}. \tag{3.38}$$

The first term vanishes because there is no time dependence and, thus, the derivative with respect to X^0 is zero. Using the list of the Christoffel symbols, given in (3.21), this equation gives

$$\frac{1}{2}e^{\nu-\lambda}\left[\nu'' + \left(\nu' - \lambda'\right)\nu'\right]$$
$$+\frac{1}{2}e^{\nu-\lambda}\left[-\left(\nu'\right)^2\right]$$
$$+\frac{1}{2}e^{\nu-\lambda}\left[\frac{(\nu')^2}{2} + \frac{\lambda'\nu'}{2} + \frac{2\nu'}{R}\right]$$
$$=$$
$$\frac{1}{2}e^{\nu-\lambda}\left[\nu'' + \frac{(\nu')^2}{2} - \frac{\lambda'\nu'}{2} + \frac{2\nu'}{R}\right]. \tag{3.39}$$

Each line corresponds to the same line as in (3.38).

(2) $\mu = \nu = 1$:

According to (3.37), he $\mathscr{R}_{11}$ component is given by

$$-\left\{\begin{matrix}\beta\\\beta1\end{matrix}\right\}_{|1} + \left\{\begin{matrix}\beta\\11\end{matrix}\right\}_{|\beta}$$
$$-\left\{\begin{matrix}\beta\\\tau1\end{matrix}\right\}\left\{\begin{matrix}\tau\\\beta1\end{matrix}\right\}$$
$$+\left\{\begin{matrix}\beta\\\tau\beta\end{matrix}\right\}\left\{\begin{matrix}\tau\\11\end{matrix}\right\}. \tag{3.40}$$

Using the list of the Christoffel symbols, given in (3.21), this equation gives

$$-\frac{\nu''+\lambda''}{2} + \frac{2}{R^2} + \frac{\lambda''}{2}$$
$$-\frac{(\nu')^2}{4} - \frac{(\lambda')^2}{4} - \frac{2}{R^2}$$
$$+\frac{\lambda'\nu'}{4} + \frac{(\lambda')^2}{4} + \frac{\lambda'}{R}$$
$$=$$
$$-\frac{1}{2}\left[\nu'' + \frac{(\nu')^2}{2} - \frac{\lambda'\nu'}{2} - \frac{2\lambda'}{R}\right]. \tag{3.41}$$

Again, each line corresponds to the same in (3.40).

(3) $\mu = \nu = 2$:
According to (3.37), the $\mathscr{R}_{22}$ component is given by

$$-\left\{\begin{matrix}\beta\\\beta 2\end{matrix}\right\}_{|2}+\left\{\begin{matrix}\beta\\22\end{matrix}\right\}_{|\beta}$$
$$-\left\{\begin{matrix}\beta\\\tau 2\end{matrix}\right\}\left\{\begin{matrix}\tau\\\beta 2\end{matrix}\right\}$$
$$+\left\{\begin{matrix}\beta\\\tau\beta\end{matrix}\right\}\left\{\begin{matrix}\tau\\22\end{matrix}\right\}. \tag{3.42}$$

Using the list of the Christoffel symbols, given in (3.21), this equation gives

$$\frac{1}{\sin^2\vartheta}-\left[Re^{-\lambda}\right]'$$
$$+2e^{-\lambda}-\cot^2\vartheta$$
$$-\left[\frac{\nu'}{2}+\frac{\lambda'}{2}+\frac{2}{R}\right]Re^{-\lambda}$$
$$=$$
$$-\left(e^{-\lambda}R\right)'-Re^{-\lambda}\frac{\nu'+\lambda'}{2}+\frac{1}{\sin^2\vartheta}-\frac{\cos^2\vartheta}{\sin^2\vartheta}$$
$$=$$
$$-\left(e^{-\lambda}R\right)'-Re^{-\lambda}\frac{\nu'+\lambda'}{2}+1, \tag{3.43}$$

where each line corresponds to the corresponding line in (3.42). In the last step also

$$\frac{1}{\sin^2\vartheta}-\frac{\cos^2\vartheta}{\sin^2\vartheta}=-\frac{\left(\cos^2\vartheta-1\right)}{\sin^2\vartheta}=1 \tag{3.44}$$

was used.

For $\mu = \nu = 3$ the steps are similar to the case with $\mu = \nu = 2$. The only difference is a factor $\sin^2\vartheta$.

We obtain from these equations and $R_{00}=\frac{1}{2}e^{\nu-\lambda}\xi_0\sigma_-$, $R_{11}=-\frac{1}{2}\xi_1\sigma_-$

$$\nu''+\frac{1}{2}\nu'^2-\frac{1}{2}\lambda'\nu'+\frac{2\nu'}{R}=\xi_0\sigma_-$$
$$\nu''+\frac{1}{2}\nu'^2-\frac{1}{2}\lambda'\nu'-\frac{2\lambda'}{R}=\xi_1\sigma_-. \tag{3.45}$$

This is the same as in [8].

Now, subtracting the second equation in (3.45) from the first one in (3.45) gives

$$\left(v' + \lambda'\right) = \frac{1}{2} R \left(\xi_0 - \xi_1\right) \sigma_-. \tag{3.46}$$

This reduces to the standard GR result, i.e. $v' + \lambda' = 0$, when ξ_0 and ξ_1 are set to zero. Equation (3.46) also implies that the σ_+ component fulfills the condition $\left(v'_+ + \lambda'_+\right) = 0$, as in standard GR, and $\left(v'_- + \lambda'_-\right) = \frac{1}{2} R_- \left(\xi_0 - \xi_1\right)$ in the σ_- component.

A repeated differentiation of (3.46) gives

$$v'' + \lambda'' = \left[\frac{1}{2} \left(\xi_0 - \xi_1\right) + \frac{R}{2} \left(\xi_0' - \xi_1'\right)\right] \sigma_-. \tag{3.47}$$

Using R_{22}, as it is given in (3.35), and comparing it to (3.26), we obtain

$$\left[Re^{-\lambda}\right]' - 1 + Re^{-\lambda} \left(\frac{v' + \lambda'}{2}\right) = \xi_2 \sigma_-. \tag{3.48}$$

Performing the derivative and rearranging gives

$$e^{-\lambda} \left[1 + \frac{Rv'}{2} - \frac{R\lambda'}{2}\right] - 1 = \xi_2 \sigma_-. \tag{3.49}$$

Substituting again (3.46) and its derivative, given by (3.47), into the left hand side of (3.45) and reordering terms leads to

$$- \left(\lambda'' - \lambda'^2 + \frac{2\lambda'}{R}\right)$$
$$+ \left[\frac{1}{2} \left(\xi_0 - \xi_1\right) + \frac{1}{2} R \left(\xi_0' - \xi_1'\right) - \frac{3}{4} \lambda' R \left(\xi_0 - \xi_1\right) + \frac{1}{8} R^2 \left(\xi_0 - \xi_1\right)^2\right] \sigma_-$$
$$= \xi_1 \sigma_- \tag{3.50}$$

For the σ_- component we have $- \left(\lambda''_- - \lambda'^2_- + \frac{2\lambda'_-}{R_-}\right) = \frac{e^{\lambda_-}}{R_-} \left(R_- e^{-\lambda_-}\right)''$ in the first term, which is verified directly. Shifting $\left[R_- e^{-\lambda_-}\right]'$ in (3.48) to one side and substituting this into $\frac{e^{\lambda_-}}{R_-} \left(R_- e^{-\lambda_-}\right)''$, we arrive at

$$- \left(\lambda''_- - \lambda'^2_- + \frac{2\lambda'_-}{R_-}\right) = \frac{e^{\lambda_-}}{R_-} \left(R_- e^{-\lambda_-}\right)''$$
$$= \frac{e^{\lambda_-}}{R_-} \left(1 + \xi_2 - \frac{R^2_-}{4} \left(\xi_0 - \xi_1\right) e^{-\lambda_-}\right)'$$
$$= \frac{e^{\lambda_-}}{R_-} \xi_2' - \frac{1}{2} \left(\xi_0 - \xi_1\right) - \frac{R_-}{4} \left(\xi_0' - \xi_1'\right) + \frac{\lambda'_- R_-}{4} \left(\xi_0 - \xi_1\right). \tag{3.51}$$

This can be inserted into the first line in (3.50), which gives

$$\frac{e^{\lambda_-}}{R_-}\xi_2' + \frac{1}{4}R_-\left(\xi_0' - \xi_1'\right) - \frac{1}{2}\lambda_-'R_-\left(\xi_0 - \xi_1\right) + \frac{1}{8}R_-^2\left(\xi_0 - \xi_1\right)^2 = \xi_1. \quad (3.52)$$

This is a general differential equation relating the ξ_μ functions, which helps to constrain them further. If a fluid model is used, these functions are related to the density and the pressure (radial and tangential) and can be transformed into the Tolman-Oppenheimer-Volkov (TOV) equation [1], derived for the Einstein equations with a matter distribution present, whose derivation will be given in Exercise 3.3:

$$\frac{p_r'}{c^2} = -\frac{\left(\varepsilon + \frac{p_r}{c^2}\right)}{R_-\left(R_- - 2M_- + 2m_{\mathrm{de}}(R_-)\right)}\left[\frac{4\pi\kappa}{c^2}R_-^3\frac{p_r}{c^2} + M_- - m_{\mathrm{de}}\right]$$
$$+ \frac{2}{R_-}\frac{\Delta p}{c^2},$$

$$(3.53)$$

with $\Delta p = p_\vartheta - p_r$. The function m_{de} is defined further below in (3.71). In case of an isotropic fluid $p_\vartheta = p_r = p$ and therefore $\Delta p = 0$. Using the equation of state $\frac{p_r}{c^2} = -\varepsilon$ (the standard equation of state for the dark energy, which however can change) the factor $\left(\varepsilon + \frac{p_r}{c^2}\right) = \left(\varepsilon + \frac{p}{c^2}\right)$ is zero and the TOV equation reduces to $p_r' = 0$, i.e. p_r is constant and due to $\frac{p_r}{c^2} \sim \varepsilon$ also ε is constant. This contradicts the requirement that ε has to vanish for large distances. One is, therefore, forced to use an anisotropic fluid. Only then Δp does not vanish. Resolving for p_ϑ one obtains a relation between p_ϑ and p_r:

$$p_\vartheta = p_r + \frac{R_-}{2}\frac{dp_r}{dR_-}. \quad (3.54)$$

When a particular behavior of p_r as a function in R_- is assumed, this determines p_ϑ. The assumption of an anisotropic fluid is not new and appears naturally in problems similar to our theory [3]. (Another example for an anisotropic fluid one encounters toward the end of this chapter related to the Reissner-Nordström solution.)

Exercise 3.3 (Derive the TOV-equation)

Solution
Using (3.30) and the definition of Ξ_μ leads to the following relations of the ξ_μ functions to the density and pressures

$$e^{-\lambda_-}(\xi_0 - \xi_1) = \frac{16\pi\kappa}{c^2}\left(\varepsilon + \frac{p_r}{c^2}\right),$$

$$e^{-\lambda_-}(\xi_0' - \xi_1') = \frac{16\pi\kappa}{c^2}\left[\lambda_-'\left(\varepsilon + \frac{p_r}{c^2}\right) + \left(\varepsilon' + \frac{p_r'}{c^2}\right)\right],$$

$$\frac{2\xi_2}{R_-^2} = \frac{8\pi\kappa}{c^2}\left(-\varepsilon + \frac{p_r}{c^2}\right),$$

$$\frac{2\xi_2'}{R_-^2} = \frac{8\pi\kappa}{c^2}\left(-\varepsilon' + \frac{p_r'}{c^2}\right) + \frac{16\pi\kappa}{c^2}\frac{1}{R_-^2}\left(-\varepsilon + \frac{p_r}{c^2}\right),$$

$$e^{-\lambda_-}\xi_1 = \frac{8\pi\kappa}{c^2}\left(-\varepsilon + 2\frac{p_\vartheta}{c^2} - \frac{p_r}{c^2}\right), \tag{3.55}$$

where, in order to obtain the second line, we first multiplied the first line with e^{λ_-}, took the derivative and then multiplied by $e^{-\lambda_-}$. For the derivative in ξ_2 we started from $2\xi_2/R_-^2$, multiplied by $R_-^2/2$, took the derivative and finally multiplied by $2/R_-^2$.

We also use the equation, derived further below,

$$e^{-\lambda_-} = 1 - \frac{2\mathcal{M}_-}{R_-} + \frac{2m_{\text{de}}(R_-)}{R_-} \tag{3.56}$$

All these expressions are substituted into (3.52), giving

$$\frac{16\pi\kappa}{c^2}\frac{1}{R_-}\left(-\varepsilon + \frac{p_r}{c^2}\right) + \frac{8\pi\kappa}{c^2}\left(-\varepsilon' + \frac{p_r'}{c^2}\right)$$
$$+\frac{1}{2}\frac{16\pi\kappa}{c^2}\left[\lambda_-'\left(\varepsilon + \frac{p_r}{c^2}\right) + \left(\varepsilon' + \frac{p_r'}{c^2}\right)\right]$$
$$-\lambda_-'\frac{16\pi\kappa}{c^2}\left(\varepsilon + \frac{p_r}{c^2}\right) + \frac{R_- e^{\lambda_-}}{4}\left(\frac{16\pi\kappa}{c^2}\right)^2\left(\varepsilon + \frac{p_r}{c^2}\right)^2$$
$$=$$
$$-\frac{2}{R_-}\frac{8\pi\kappa}{c^2}\left(\varepsilon + \frac{p_r}{c^2}\right) + \frac{2}{R_-}\frac{8\pi\kappa}{c^2}2\frac{p_\vartheta}{c^2}. \tag{3.57}$$

As one can readily notes, the derivative in the density cancels. Resolving for $\frac{p_r'}{c^2}$ we obtain

$$\frac{p_r'}{c^2} = \frac{2}{R_-}\frac{(p_\vartheta - p_r)}{c^2} + \left(\varepsilon + \frac{p_r}{c^2}\right)\frac{\lambda_-'}{2} - \frac{8\pi\kappa}{c^2}\frac{R_-}{2}\left(\varepsilon + \frac{p_r}{c^2}\right)^2. \tag{3.58}$$

Next, the λ_-' is determined. For that we use the relation of $e^{-\lambda_-}$, as given in (3.56), deriving $e^{-\lambda_-}$ with respect to R_-, which gives

$$-\lambda_- e^{-\lambda_-} = \frac{2\mathcal{M}_-}{R_-^2} - \frac{2m_{\text{de}}}{R_-^2} + \frac{2m_{\text{de}}'}{R_-}. \tag{3.59}$$

Resolving for λ'_- gives

$$\lambda_- = -\frac{e^{-\lambda_-}}{R_-^2}\left(2\mathcal{M}_- - 2m_{\mathrm{de}} + 2R_-m'_{\mathrm{de}}\right).$$ (3.60)

Substituting this into (3.58) gives, with $\Delta p = p_\vartheta - p_r$,

$$\frac{p'}{c^2} = \frac{2}{R_-}\frac{\Delta p}{c^2} - \left(\varepsilon + \frac{p_r}{c^2}\right)\frac{e^{\lambda_-}}{R_-^2}\left[\mathcal{M}_- - m_{\mathrm{de}} + R_-m'_{\mathrm{de}} + \frac{4\pi\kappa}{c^2}R_-^3\left(\varepsilon + \frac{p_r}{c^2}\right)\right],$$ (3.61)

Using that

$$m'_{\mathrm{de}} = -\frac{4\pi\kappa}{c^2}R_-^2\varepsilon$$ (3.62)

cancels the density dependence in the parenthesis and leads to the following expression for the derivative of the radial pressure:

$$\frac{p'_r}{c^2} = -\frac{\left(\varepsilon + \frac{p_r}{c^2}\right)}{R_-\left(R_- - 2\mathcal{M}_- + 2m_{\mathrm{de}}\right)}\left[\frac{4\pi\kappa}{c^2}R_-^2\frac{p_r}{c^2} + \mathcal{M}_- - m_{\mathrm{de}}\right] + \frac{2}{R_-}\frac{\Delta p}{c^2},$$ (3.63)

which is nothing but the TOV equation for an anisotropic fluid. The TOV equation for an isotropic fluid is obtained as a special case, where $p_\vartheta = p_r$.

Using the derivative of $R_- e^{-\lambda_-}$ as given in (3.48), integrating it and setting the integration constant equal to $-2\mathcal{M}_-$, we obtain

$$R_- e^{-\lambda_-} = R_- - 2\mathcal{M}_- + \int \xi_2 dR_- - \frac{1}{4}\int e^{-\lambda_-}R_-^2\left(\xi_0 - \xi_1\right)dR_-.$$ (3.64)

The sum of the terms within the integral is nothing but $\varXi_0 R_-^2$, with $\varXi_0$ as given in (3.30) and is simply proportional to the energy density.

Now, we can use the connection of the ξ_μ functions in terms of the density ε and pressures p_r and p_ϑ, given in (3.34), with the result

$$R_- e^{-\lambda_-} = R_- - 2\mathcal{M}_- - \frac{8\pi\kappa}{c^2}\int R_-^2\varepsilon dR_-.$$ (3.65)

This general expression is always valid, i.e., the additional contribution depends on ε only. It is the same for an isotropic and an anisotropic fluid!

Exercise 3.4 (**Equations** (3.64) **and** (3.65))

Problem. Derive (3.64) and (3.65).

Solution. Starting from (3.48), resolving for $\left[R_-e^{-\lambda_-}\right]'$, we get

$$\left[R_-e^{-\lambda_-}\right]' = \xi_2 + 1 - R_-e^{-\lambda_-}\left(\frac{\nu' + \lambda'}{2}\right). \tag{3.66}$$

Using (3.46) and integrating yields

$$R_-e^{-\lambda_-} = R_- - 2\mathcal{M}_- + \int \xi_2 dR_- - \frac{1}{4}\int e^{-\lambda_-}R_-^2\left(\xi_0 - \xi_1\right)dR_-, \tag{3.67}$$

where $\mathcal{M}_-$ is an integration constant. This is (3.64), where the integrand of the integral over R_- is

$$\xi_2 - \frac{R_-^2}{4}e^{-\lambda_-}\left(\xi_0 - \xi_1\right) = R_-^2 \Xi_0. \tag{3.68}$$

We have used the first equation in (3.30), multiplied by R_-^2. Because $\Xi_0 = -\frac{8\pi\kappa}{c^2}\varepsilon$, this integrand is nothing but $-\frac{8\pi\kappa}{c^2}R_-^2\varepsilon$ as it appears in (3.65).

The metric component $e^{-\nu_-}$, has a more involved expression: Using (3.46), we have $\nu'_- = -\lambda'_- + \frac{R_-}{2}\left(\xi_0 - \xi_1\right)$. Integrating over R_- gives $\nu_- = -\lambda_- + \frac{1}{2}\int R_- \left(\xi_0 - \xi_1\right)dR_-$. Elevating this to the power of the exponential yields

$$e^{\nu_-} = e^{-\lambda_-}e^{\frac{1}{2}\int R_-(\xi_0-\xi_1)dR_-} \tag{3.69}$$

From the second equation in (3.31) we have $\frac{e^{-\lambda_-}}{2}\left(\xi_0 - \xi_1\right) = \Xi_1 - \Xi_0 = \frac{8\pi\kappa}{c^2}\left(\frac{p_r}{c^2} + \varepsilon\right)$ (see (3.33)). Thus, the latter equation results into

$$e^{-\lambda_-}e^{\frac{8\pi\kappa}{c^2}\int R_-e^{\lambda_-}\left[\left(\frac{p_r}{c^2}\right)+\varepsilon\right]dR_-} = e^{-\lambda_-}e^{f_-}. \tag{3.70}$$

The factor e^{f_-} is a positive function in R_- and apart from the density ε it depends only on the radial pressure.

We now proceed in rewriting (3.65) and define

$$-\frac{8\pi\kappa}{c^2}\int R_-^2\varepsilon dR_- = \frac{2\kappa}{c^2}M_{\mathrm{de}}(R_-)$$
$$= 2m_{\mathrm{de}}(R_-), \tag{3.71}$$

which gives the definition of m_{de} used further below. $M_{de}(R_-)$ is the accumulated mass of the dark energy. In order that the contribution $\frac{2m_{de}}{R_-}$ to the metric component g_{00}^- is at least proportional $1/R_-^3$ ($m_{de} \sim M_{de}$), the M_{de} has to be at least proportional to $1/R_-^2$. This is only achieved if the density is proportional at least to $1/R_-^5$. Thus, the integrand $-R_-^2\varepsilon$ is proportional to $-1/R_-^3$. Integrating gives an expression proportional to $+1/R_-^2$. As one can see, the minus sign in (3.71) guarantees, for positive ε and the upper integration limit larger than the lower one, that the dark energy mass is positive. This may change when the integration limits are different and/or the energy density has different signs, which will be explored in Chap. 6, where neutron stars are investigated.

The metric element g_-^{11} is equal to $e^{-\lambda_-}$, which is obtained from (3.65) and substituting it in the integral by (3.71). The g_-^{11} is finally given by

$$e^{-\lambda_-} = 1 - \frac{2\mathcal{M}_-}{R} + \frac{2m_{de}(R_-)}{R_-}. \tag{3.72}$$

In (3.70) the $g_{00}^- = -e^{\nu_-}$ was related to $e^{-\lambda_-}$, which is the g_-^{11} metric component. In general, an additional factor may appear, abbreviated by e^{f_-}. With this and (3.72), the g_{00}^- component acquires the structure

$$g_{00} = -\left(1 - \frac{2\mathcal{M}_-}{R_-} + \frac{2m_{de}}{R_-}\right)e^{f_-}. \tag{3.73}$$

As was shown above, for an anisotropic fluid and the equation of state $\frac{p_r}{c^2} = -\varepsilon$, the function f_- is zero and a simple expression for g_{00}^- is obtained.

In what follows, the final pc-Schwarzschild solution is constructed:

Up to here we obtained an analytic solution for the σ_- component of the metric. The one for the σ_+ component is identical to the one derived by e.g. Adler et al. [1], though having used a different signature. In the σ_- component, the additional contribution proportional to the function $\Omega_- = 2m_{de}(R_-)$ appears, where we stressed that it is a function in R_-. However such a function does *not* appear in the σ_+ component. In order to rewrite the σ_+ component in a form similar to the one in the σ_- component, we use the prescription enlisted in (3.4)–(3.8) and introduce the definitions $\Omega_+ = 0$ and $f_+ = 0$. With the help of this, both components can be written as

$$\left(g_{\mu\nu}^\pm\right) = \begin{pmatrix} -\left(1 - \frac{2\mathcal{M}_\pm}{R_\pm} + \frac{\Omega_\pm}{R_\pm}\right)e^{f_\pm} & 0 & 0 & 0 \\ 0 & \left(1 - \frac{2\mathcal{M}_\pm}{R_\pm} + \frac{\Omega_\pm}{R_\pm}\right)^{-1} & 0 & 0 \\ 0 & 0 & R_\pm^2 & 0 \\ 0 & 0 & 0 & R_\pm^2\sin^2\vartheta \end{pmatrix}. \tag{3.74}$$

Note, the metric tensor has now the same functional form in both the σ_- and σ_+ component. The following notation was used

$$
\begin{aligned}
R_\pm &= r \pm l\dot{r} \\
\mathcal{M} &= \mathcal{M}_+ \sigma_+ + \mathcal{M}_- \sigma_- \\
\mathcal{M}_\pm &= m \\
\Omega &= 2m_{\text{de}}\sigma_- = \Omega_+ \sigma_+ + \Omega_- \sigma_- \\
\Omega_+ &= 0, \quad \Omega_- = 2m_{\text{de}}(R_-) = \frac{B_-}{R_-^2}.
\end{aligned}
\tag{3.75}
$$

$\mathcal{M}_\pm = m$, which is a consequence of the boundary condition that standard GR should be obtained in the limit of large distances R_-. The pseudo-real elements of the parameters are

$$
\mathcal{M}_R = m \,, \quad B_R = \frac{1}{2}(B_+ + B_-) = \frac{B_-}{2}.
\tag{3.76}
$$

Because now the metric tensors in both σ-component have the same functional form, the total pseudo-complex metric can be written as $g_{\mu\nu}(B, R) = g_{\mu\nu}(B_+, R_+)\sigma_+ + g_{\mu\nu}(B_-, f_-, R_-)\sigma_-$, which gives

$$
\left(g_{\mu\nu}\right) = \begin{pmatrix}
-\left(1 - \frac{2\mathcal{M}}{R} + \frac{B}{R^3}\right)e^f & 0 & 0 & 0 \\
0 & \left(1 - \frac{2\mathcal{M}}{R} + \frac{B}{R^3}\right)^{-1} & 0 & 0 \\
0 & 0 & R^2 & 0 \\
0 & 0 & 0 & R^2\sin^2\vartheta
\end{pmatrix},
\tag{3.77}
$$

with $f_+ = f_- = 0$ and therefore $e^f = 1$. The projected metric, following our prescription, is

$$
\left(g_{\mu\nu}(r)\right) = \begin{pmatrix}
-\left(1 - \frac{2m}{r} + \frac{B_-}{2r^3}\right) & 0 & 0 & 0 \\
0 & \left(1 - \frac{2m}{r} + \frac{B_-}{2r^3}\right)^{-1} & 0 & 0 \\
0 & 0 & r^2 & 0 \\
0 & 0 & 0 & r^2\sin^2\vartheta
\end{pmatrix}.
\tag{3.78}
$$

The real length element squared is finally given by (setting $B_- = B$)

$$d\omega^2 = -\left(1 - \frac{2m}{r} + \frac{B}{2r^3}\right)(dx^0)^2 + \left(1 - \frac{2m}{r} + \frac{B}{2r^3}\right)^{-1}(dr)^2$$
$$+ r^2\left[(d\vartheta)^2 + \sin^2\vartheta\,(d\varphi)^2\right].$$

$$(3.79)$$

In [8] we imposed the condition $-g_{00}(r) > 0$, so that the signature for the time stays the same. The condition $-g_{00}(r) > 0$ then is

$$1 - \frac{2m}{r} + \frac{B}{2r^3} > 0. \tag{3.80}$$

To find a value for B which satisfies this condition for all $r > 0$ we will have a look at the extremal value of g_{00}. As we know from the limiting behavior of g_{00} ($-g_{00} \to 1$ for $r \to \infty$ and $-g_{00} \to +\infty$ for $r \to 0$) its extremal value will be a minimum. A quick calculation gives $r = \left(\frac{3}{4}\frac{B}{m}\right)^{1/2}$ for the value of the minimum of g_{00}. Inserting this into (3.80) yields

$$B > \frac{64}{27}m^3. \tag{3.81}$$

Exercise 3.5 (Limit for the parameter B)

Problem. Proof (3.81).

Solution. Define the function

$$f(r) = 1 - \frac{2m}{r} + \frac{B}{2r^3}. \tag{3.82}$$

The first derivative with respect to r is

$$f' = \frac{2m}{r^2} - \frac{3B}{2r^4} = 0, \tag{3.83}$$

which is set to zero, in order to find the minimum. Multiplying with r^4 and resolving for r, changing r to r_0 in order to indicate that the solution is related to a minimum, the solution of this equation is

$$r_0 = \sqrt{\frac{3B}{4m}}. \qquad (3.84)$$

In order to obtain the value of the function $f(r_0)$, (3.84) is substituted into (3.82), giving

$$1 - 2m\sqrt{\frac{4m}{3B}} + \frac{B}{2}\left(\frac{4m}{3B}\right)\sqrt{\frac{4m}{3B}} = 1 - \frac{4}{3}m\sqrt{\frac{4m}{3B}} > 0, \qquad (3.85)$$

where we include the requirement that this function $f(r) = -g_{00}$ is always positive. If this is the case, the g_{00} is never zero and, thus, *no event horizon appears*.

Resolving (3.85) for B gives the condition

$$B > \frac{64}{27}m^3. \qquad (3.86)$$

The conclusions from [8] is that the redshift in the pc-Schwarzschild solution is first increased until the minimum of g_{00} is reached, from which on it decreases again until it turns into a blueshift. Because the potential is proportional to the square root of $g_{00}(r)$ [9], this indicates a minimum in the potential, which is repulsive for lower radial distances r (see Fig. 3.1). Our interpretation of this finding is that the collapse

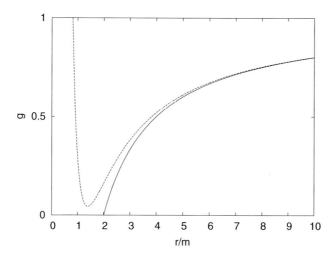

Fig. 3.1 The function $g = \sqrt{-g_{00}}$, which is proportional to the effective potential with angular momentum equal to zero [9]. The *solid line* is the Schwarzschild solution and the *dashed line* the pc-Schwarzschild solution, with $B = \frac{70}{27}m^3$. The x-axis is given in units of m, i.e., $2m$ corresponds to the Schwarzschild radius. Note, that in the pc-case a repulsive core appears which is responsible for halting the collapse of a large mass

of a star is halted latest at the minimum and it can not contract to a singularity. The star probably still oscillates around this minimum, which should be eventually observable.

The result has to be taken with care for values of r smaller than r_0. This is because we assumed to be *outside* the mass distribution of the star. As soon as the surface of the star is reached, which happens near the minimum of g_{00}, the energy-momentum tensor for the baryonic mass distribution has to be included. Thus, the behavior, including its pole toward $r = 0$ is only of academic interest and should be modified such that at $r = 0$ the $-g_{00} = |g_{00}|$ component of the metric approaches ≤ 1. This should be the correct limit, because within the star mass is present, not taken into account yet, and at the center of the star the gravitational attraction is lower. We recommend to consult Chap. 6 on neutron stars for further clarification.

3.3 The pc-Kerr Solution

In the last chapter we studied the solution of the Einstein equations for a spherical symmetric body. This was the first non-trivial solution found by K. Schwarzschild (please, see the references [1, 9]). Realistic stars are rotating, but for a long time no solution was found until R. Kerr (please, see the references [1, 9]) encountered in 1963 a solution for rotating bodies. This is very important, because the large mass concentrations at the center of nearly every galaxy, associated to so-called black holes, must have a high rotational frequency. This is expected due to the fact that a large mass collapses with an initial angular momentum and similar to the effect of an ice skater, the rotational frequency increases when the mass is more concentrated toward the rotational axis.

The derivation of the solution is done in complete analogy as given in the literature cited in the text. The difference is the additional presence of an energy-momentum tensor (for the dark energy), which will introduce some complications. The standard solution in GR is obtained by setting the parameter $B = bm^3$, describing the coupling of the star's mass to the vacuum fluctuations, to zero. As a useful byproduct, also the solution in standard GR with the presence of a mass distribution is obtained, when the additional energy-momentum tensor is interpreted as the one for mass.

As we will see, specific assumptions have to be made in order to solve the Einstein equations. i.e. the solution is not the most general one. However, the solution obtained describes a rotational body with the presence of a dark energy and is general enough to describe highly rotational large mass concentrations. The solution will play a central role in Chap. 5, where we discuss experimental predictions of pc-GR applied to rotating large masses as we encounter in the center of our galaxy.

Some intermediate steps of the calculations can be found in the appendix of [10] and in [2].

To find a pseudo-complex Kerr solution is not at all trivial. We will follow an ansatz proposed by Carter [11, 12]. The Klein-Gordon-Equation

$$\frac{1}{\Psi} \frac{\partial}{\partial x^{\alpha}} \left(\sqrt{-g} g^{\alpha\beta} \frac{\partial \Psi}{\partial x^{\beta}} \right) - m_0^2 \sqrt{-g} = 0 \qquad (3.87)$$

is required to be separable. For the structure of the metric tensor this implies

$$g_{\mu\nu} = \frac{1}{Z} \begin{pmatrix} -\Delta_r C_\mu^2 + \Delta_\mu C_r^2 & 0 & 0 & -\Delta_\mu C_r Z_r + \Delta_r C_\mu Z_\mu \\ 0 & \frac{Z^2}{\Delta_r} & 0 & 0 \\ 0 & 0 & \frac{Z^2}{\Delta_\mu} & 0 \\ -\Delta_\mu C_r Z_r + \Delta_r C_\mu Z_\mu & 0 & 0 & -\Delta_r Z_\mu^2 + \Delta_\mu Z_r^2 \end{pmatrix}, \quad (3.88)$$

where, Δ_r and Z_r are functions of the variable r while Δ_μ and Z_μ are functions of $\mu = \cos\vartheta$. C_μ and C_r are constant factors, which are determined later. The definition for Z is $Z = Z_r C_\mu - Z_\mu C_r$. The metric shows a high symmetry, but it is still quite involved to compute the Einstein equation. A manageable solution can be obtained using differential geometry, introduced by Cartan [13] into the problem under discussion. For a detailed explanation the reader is referred to [2, 9, 14, 15]. The Einstein tensor ($G_\nu^\mu = R_\nu^\mu - \frac{1}{2} g_\nu^\mu R$) obtained is given by

$$G^0{}_0 = \frac{1}{2Z} \Delta_{\mu|\mu\mu} + \frac{1}{Z^2} \Delta_r Z_{r|rr} + \frac{a^2}{4Z^3} \Delta_\mu \left(Z_{\mu|\mu}{}^2 + Z_{r|r}{}^2 \right)$$
$$- \frac{3}{4Z^3} \Delta_r \left(Z_{\mu|\mu}{}^2 + Z_{r|r}{}^2 \right) + \frac{a}{2Z^2} \Delta_{\mu|\mu} Z_{\mu|\mu} + \frac{1}{2Z^2} \Delta_{r|r} Z_{r|r}$$

$$G^0{}_3 = -\frac{1}{2Z^2} \sqrt{\Delta_r \Delta_\mu} \left(a Z_{r|rr} + Z_{\mu|\mu\mu} \right)$$

$$G^1{}_1 = \frac{1}{2Z} \Delta_{\mu|\mu\mu} + \frac{a^2}{4Z^3} \Delta_\mu \left(Z_{\mu|\mu}{}^2 + Z_{r|r}{}^2 \right) - \frac{1}{4Z^3} \Delta_r \left(Z_{\mu|\mu}{}^2 + Z_{r|r}{}^2 \right)$$
$$+ \frac{a}{2Z^2} \Delta_{\mu|\mu} Z_{\mu|\mu} + \frac{1}{2Z^2} \Delta_{r|r} Z_{r|r}$$

$$G^2{}_2 = \frac{1}{2Z} \Delta_{r|rr} - \frac{a^2}{4Z^3} \Delta_\mu \left(Z_{\mu|\mu}{}^2 + Z_{r|r}{}^2 \right) + \frac{1}{4Z^3} \Delta_r \left(Z_{\mu|\mu}{}^2 + Z_{r|r}{}^2 \right)$$
$$- \frac{a}{2Z^2} \Delta_{\mu|\mu} Z_{\mu|\mu} - \frac{1}{2Z^2} \Delta_{r|r} Z_{r|r}$$

$$G^3{}_3 = \frac{1}{2Z} \Delta_{r|rr} - \frac{a}{Z^2} \Delta_\mu Z_{\mu|\mu\mu} - \frac{3a^2}{4Z^3} \Delta_\mu \left(Z_{\mu|\mu}{}^2 + Z_{r|r}{}^2 \right)$$
$$+ \frac{1}{4Z^3} \Delta_r \left(Z_{\mu|\mu}{}^2 + Z_{r|r}{}^2 \right) - \frac{a}{2Z^2} \Delta_{\mu|\mu} Z_{\mu|\mu} - \frac{1}{2Z^2} \Delta_{r|r} Z_{r|r}, \quad (3.89)$$

where the constant factors are chosen as $C_r = a$, $C_\mu = 1$ [11, 12, 16] and the subscript $_{|\mu,r}$ stand for the derivative with respect to μ, r respectively. The calculations are similar to the pc- Schwarzschild case. With the variational principle modified, only the σ_- component of the equation needs to be considered, as the σ_+ part of it is identical to the classical Einstein equation. To solve the Einstein equation $G^\nu{}_\mu = \Xi^\nu_\mu \sigma_-$, we consider similar combinations of (3.89) as shown by Plebánski and Krasiński [12]:

$$-\frac{1}{2Z^2}\sqrt{\Delta_{R_-}\Delta_{\mu_-}}\,(aZ_{R_-|R_-R_-}+Z_{\mu_-|\mu_-\mu_-})=\Xi^0{}_3$$

$$\frac{1}{2Z}(\Delta_{\mu_-|\mu_-\mu_-}+\Delta_{R_-|R_-R_-})=\Xi^1{}_1+\Xi^2{}_2$$

$$\frac{a}{Z^2}\Delta_{\mu_-}Z_{\mu_-|\mu_-\mu_-}+\frac{a^2}{2Z^3}\Delta_{\mu_-}(Z^2_{\mu_-|\mu_-}+Z^2_{R_-|R_-})=\Xi^2{}_2-\Xi^3{}_3$$

$$\frac{1}{Z^2}\Delta_{R_-}Z_{R_-|R_-R_-}-\frac{1}{2Z^3}\Delta_{R_-}(Z^2_{\mu_-|\mu_-}+Z^2_{R_-|R_-})=\Xi^0{}_0-\Xi^1{}_1$$

$$\frac{1}{2Z}\Delta_{R_-|R_-R_-}-\frac{a^2}{4Z^3}\Delta_{\mu_-}\left(Z_{\mu_-|\mu_-}{}^2+Z_{R_-|R_-}{}^2\right)-\frac{a}{2Z^2}\Delta_{\mu_-|\mu_-}Z_{\mu_-|\mu_-}$$
$$+\frac{1}{4Z^3}\Delta_{R_-}\left(Z_{\mu_-|\mu_-}{}^2+Z_{R_-|R_-}{}^2\right)-\frac{1}{2Z^2}\Delta_{R_-|R_-}Z_{R_-|R_-}=\Xi^2{}_2. \tag{3.90}$$

The Ξ^ν_μ are first considered as arbitrary functions in R_- and μ_-. This allows us to choose them properly, so that the equations in (3.90) can be solved. The first step consists in setting $\Xi^3_0=0$ and thus the first line in (3.90) becomes

$$aZ_{R_-|R_-R_-}+Z_{\mu_-|\mu_-\mu_-}=0. \tag{3.91}$$

Choosing $\Xi^3_0\neq0$ would not allow an analytic solution, i.e., the assumption $\Xi^3_0=0$ *is for convenience*. (3.91) is formally identical to the classical case [12]. We have a sum of two functions of different variables equal to a constant. Both have to be constant and one can conclude that

$$Z_{R_-}=CR^2_-+C_1R_-+C_2\quad\text{and}\quad Z_{\mu_-}=-aC\mu^2_-+C_3\mu_-+C_4. \tag{3.92}$$

Without further assumptions, no solution can be found yet. However, using as an ad hoc choice $\Xi^2_2=\Xi^3_3$, a solution can be found, as shown in what follows. Here, identifying the index 2 with ϑ and 3 with φ, a symmetry is assumed, that the Ξ^2_2 has the same form as Ξ^3_3, which allows still a quite general solution. Also with the third equation in (3.90) we arrive, after a trivial calculation, at

$$C_4=\frac{C_2}{a}-\frac{C^2_1+C^2_3}{4aC}. \tag{3.93}$$

Inserted into (3.91) we can observe, that the transformation $\mu_-=\mu'_-+\frac{C_3}{2aC}$ together with a redefinition $C_2=aC'_2+\frac{C^2_1}{4C}$ has the same effect as if we would choose $C_3=0$ [12]. Thus, we are left with

$$Z_{R_-}=C\left(R_-+\frac{C_1}{2C}\right)^2+aC'_2,\quad Z_{\mu_-}=-aC\mu^2_-+C'_2. \tag{3.94}$$

A different transformation for the variable R_- yields the same as if we would set $C_1 = 0$.

Exercise 3.6 (Equations (3.92) **and** (3.94))

Problem. Show that (3.94) follows from (3.92).

Solution. The term (3.94) is written explicitly, i.e.

$$Z_{R_-} = C\left(R_-^2 + \frac{C_1}{C}R_- + \frac{C_1^2}{4C^2}\right) + aC_2'$$

$$= CR_-^2 + C_1 R_- + \frac{C_1^2}{4C} + aC_2'. \tag{3.95}$$

Identifying $\frac{C_1^2}{4C} + aC_2'$ with C_2 in (3.92) shows the equivalence. This allows to rescale $\left(R_- + \frac{C_1}{2C}\right)$ to a new R_-. In other words, instead of using an arbitrary C_1, it is save to assume in (3.92) that $C_1 = 0$

The Z_{μ_-} in (3.94) is obtained from (3.92) in the same way, using the equivalent rescaling for μ_- as for R_- in the former example, i.e., one can set C_3 to zero and just rename the parameter C_4 to C_2'.

As the factor $Z = Z_{R_-} - aZ_{\mu_-}$ is independent of C_2' we can choose $C_2' = aC$, just as in the classical case [12]. The final step consists in setting $C = 1$, redefining Δ_{μ_-} and Δ_{R_-}. Therefore, we obtain the functions

$$Z_{R_-} = R_-^2 + a^2, \quad Z_{\mu_-} = a(1 - \mu_-^2), \tag{3.96}$$

which again are formally identical to the classical solution [11, 12].

We shift our attention to the second equation of (3.90) with the assumption $\Xi_1^1 + \Xi_2^2 = \frac{1}{2Z}\sum_{n=3}^{\infty}\frac{\tilde{B}_n}{R_-^n}$ (the right hand side simulates the contribution of $T_\Lambda^{\mu\nu}$ of the "dark energy"), which yields

$$\Delta_{R_-|R_-R_-} + \Delta_{\mu_-|\mu_-\mu_-} - \sum_{n=3}^{\infty}\frac{\tilde{B}_n}{R_-^n} = 0. \tag{3.97}$$

Again we have two functions in different variables and they have to be equal. This leads to

$$\Delta_{R_-} = ER_-^2 - 2\mathcal{M}_-R_- + E_2 + \sum_{n=3}^{\infty} \frac{1}{(n-1)(n-2)} \frac{\tilde{B}_n}{R_-^{n-2}},$$

$$\Delta_{\mu_-} = -E\mu_-^2 + E_3\mu_- + E_4. \tag{3.98}$$

Inserting this and (3.96) into the last equation of (3.90) gives, after some algebra,

$$\sum_{n=3}^{\infty} \left(\frac{\tilde{B}_n}{R_-^{n-2}} \left(\frac{1}{n-2} + \frac{1}{2} \right) + \frac{\tilde{B}_n a^2 \mu_-^2}{2R_-^n} \right) + (E_2 - E_4 a^2) = Z^2 \Xi^2_2. \tag{3.99}$$

If we chose

$$\Xi^2_2 = \frac{1}{Z^2} \sum_{n=3}^{\infty} \left(\frac{\tilde{B}}{R_-^{n-2}} \left(\frac{1}{n-2} + \frac{1}{2} \right) + \frac{\tilde{B} a^2 \mu_-^2}{2R_-^n} \right), \tag{3.100}$$

the previous equation can be fulfilled while maintaining the condition $(E_2 - E_4 a^2) = 0$, as in the classical case.

In order to determine the remaining constants in Δ_{R_-} and Δ_{μ_-} we will proceed analogously to Plebánski and Krasiński [12]. First, we set $E_3 = 0$, otherwise one would obtain a term proportional to $\mu_- = \cos\vartheta$, which violates the symmetry with respect to a reflection on the equatorial plane. To avoid a coordinate singularity at the poles, we set $E = 1$. Finally we choose $E_4 = 1$ to get the correct Schwarzschild metric in the limit $a \to 0$. This leaves us with

$$Z_{R_-} = R_-^2 + a^2, \quad Z_{\mu_-} = a(1 - \mu_-^2),$$

$$\Delta_{R_-} = R_-^2 - 2\mathcal{M}_-R_- + a^2 + \sum_{n=3}^{\infty} \frac{1}{(n-1)(n-2)} \frac{\tilde{B}_n}{R_-^{n-2}},$$

$$\Delta_{\mu_-} = 1 - \mu_-^2, \quad Z = Z_{R_-} - aZ_{\mu_-} = R_-^2 + a^2\mu_-^2. \tag{3.101}$$

With this inserted into (3.88) and, together with $\mu = \cos\vartheta$, we get the σ_--part of the metric as

$$\bar{g}_{00} = -\frac{R_-^2 - 2\mathcal{M}_-R_- + a^2\cos^2\vartheta_- + \sum_{n=3}^{\infty} \frac{1}{(n-1)(n-2)} \frac{\tilde{B}_n}{R_-^{n-2}}}{R_-^2 + a^2\cos^2\vartheta_-}$$

$$\bar{g}_{11} = \frac{R_-^2 + a^2\cos^2\vartheta_-}{R_-^2 - 2\mathcal{M}_-R_- + a^2 + \sum_{n=3}^{\infty} \frac{1}{(n-1)(n-2)} \frac{\tilde{B}_n}{R_-^{n-2}}}$$

$$\bar{g}_{22} = R_-^2 + a^2\cos^2\vartheta_-$$

$$\bar{g}_{33} = (R_-^2 + a^2)\sin^2\vartheta_- + \frac{a^2\sin^4\vartheta_- \left(2\mathcal{M}_-R_- - \sum_{n=3}^{\infty} \frac{1}{(n-1)(n-2)} \frac{\tilde{B}_n}{R_-^{n-2}} \right)}{R_-^2 + a^2\cos^2\vartheta_-}$$

$$g_{\bar{0}3} = \frac{-a \sin^2 \vartheta_- - 2\mathcal{M}_- R_- + a \sum_{n=3}^{\infty} \frac{1}{(n-1)(n-2)} \frac{\tilde{B}_n}{R_-^{n-2}} \sin^2 \vartheta_-}{R_-^2 + a^2 \cos^2 \vartheta_-}. \tag{3.102}$$

Note, that in spite of all assumptions made, (3.102) represents a new Kerr solution also in standard GR with a special T_ν^μ tensor.

The σ_+-component matches the classical Kerr solution. Finally, projecting the pc-metric on its real part, as described in the pc-Schwarzschild case and in the introductory chapter on pc-GR, yields the metric

$$g_{00}^{\text{Re}} = -\frac{r^2 - 2mr + a^2 \cos^2 \vartheta + \sum_{n=3}^{\infty} \frac{1}{(n-1)(n-2)} \frac{\tilde{B}_n}{2r^{n-2}}}{r^2 + a^2 \cos^2 \vartheta}$$

$$g_{11}^{\text{Re}} = \frac{r^2 + a^2 \cos^2 \vartheta}{r^2 - 2mr + a^2 + \sum_{n=3}^{\infty} \frac{1}{(n-1)(n-2)} \frac{\tilde{B}_n}{2r^{n-2}}}$$

$$g_{22}^{\text{Re}} = r^2 + a^2 \cos^2 \vartheta$$

$$g_{33}^{\text{Re}} = (r^2 + a^2) \sin^2 \vartheta + \frac{a^2 \sin^4 \vartheta \left(2mr - \sum_{n=3}^{\infty} \frac{1}{(n-1)(n-2)} \frac{\tilde{B}_n}{2r^{n-2}}\right)}{r^2 + a^2 \cos^2 \vartheta}$$

$$g_{03}^{\text{Re}} = \frac{-a \sin^2 \vartheta \, 2mr + a \sum_{n=3}^{\infty} \frac{1}{(n-1)(n-2)} \frac{\tilde{B}_n}{2r^{n-2}} \sin^2 \vartheta}{r^2 + a^2 \cos^2 \vartheta}. \tag{3.103}$$

In what follows, we only consider the case $n = 3$, i.e., the metric

$$g_{00}^{\text{Re}} = -\frac{r^2 - 2mr + a^2 \cos^2 \vartheta + \frac{B}{2r}}{r^2 + a^2 \cos^2 \vartheta}$$

$$g_{11}^{\text{Re}} = \frac{r^2 + a^2 \cos^2 \vartheta}{r^2 - 2mr + a^2 + \frac{B}{2r}}$$

$$g_{22}^{\text{Re}} = -r^2 + a^2 \cos^2 \vartheta$$

$$g_{33}^{\text{Re}} = (r^2 + a^2) \sin^2 \vartheta + \frac{a^2 \sin^4 \vartheta \left(2mr - \frac{B}{2r}\right)}{r^2 + a^2 \cos^2 \vartheta}$$

$$g_{03}^{\text{Re}} = \frac{-a \sin^2 \vartheta \, 2mr + a \frac{B}{2r} \sin^2 \vartheta}{r^2 + a^2 \cos^2 \vartheta}. \tag{3.104}$$

Because (3.104) represents the pseudo-complex equivalent to the Kerr solution, it is important to ask whether one can still identify the parameter a with the angular momentum J. To do so, we follow Adler et al. [1] and expand the line element given by (3.104) linear in a

$$ds^2 = -\left(1 - \frac{2m}{r} + \frac{B}{2r^3}\right)dt^2 + \frac{1}{1 - \frac{2m}{r} + \frac{B}{2r^3}}dr^2 - r^2 d\vartheta^2 - r^2 \sin^2\vartheta d\varphi^2$$

$$+2a\sin^2\vartheta\left(-\frac{2m}{r} + \frac{B}{2r^3}\right)d\varphi dt. \tag{3.105}$$

This expansion represents the limit of a slowly rotating body. Next we expand (3.105) linear in $\frac{1}{r}$, which is the limit for large distances. The line element takes the form

$$ds^2 = -\left(1 - \frac{2m}{r}\right)dt^2 + \left(1 + \frac{2m}{r}\right)dr^2 + r^2 d\vartheta^2 - r^2 \sin^2\vartheta d\varphi^2$$

$$+2a\sin^2\vartheta\frac{2m}{r}d\varphi dt. \tag{3.106}$$

In what follows, we will focus on the term proportional to $d\varphi dt$.

$$+ 2a\sin^2\vartheta\frac{2m}{r}d\varphi dt. \tag{3.107}$$

A comparison of this term with the metric far from a stationary rotating source shows that the parameter m denotes the mass and a is proportional to the intrinsic angular momentum [9]

$$ds^2 = -\left(1 - \frac{2m}{\rho}\right)dt^2 + \left(1 + \frac{2m}{\rho}\right)\left[d\rho^2 + \rho^2\left(d\vartheta^2 + \sin^2\vartheta d\varphi^2\right)\right]$$

$$+\frac{4\kappa J}{c^3\rho}\sin^2\vartheta d\varphi dt. \tag{3.108}$$

Note, that for large distances the two radial variables r and ρ (isotropic coordinates, used by Lense and Thirring) coincide. This finally yields the same connection between a and the angular momentum J as in the classical case

$$a = \frac{\kappa J}{mc^3}. \tag{3.109}$$

Obviously, the parameter a can still be identified with the angular momentum of the source.

The classical Kerr solution has particular hypersurfaces which are of great physical interest. One of these is the same as in the classical Schwarzschild solution: In the orbital plane the radius of this sphere is at $r = 2m$, a sphere with infinite red shift. For corrections in the metric proportional to $\frac{B}{2r^3}$, the infinite redshift surface and the singularity at the center vanish for $B > (4/3m)^3$ (see (3.81)). We will now investigate the influence of the additional term proportional to B on the existence of an event horizon for the Kerr metric.

As shown in [1] surfaces corresponding to $g_{00} = 0$ can be passed in both directions by an observer (except at the poles), e.g., these surfaces are no event horizons.

The property of a surface to be an event horizon is determined by the norm of its normal vector n_α. Only if it is positive, physical observers can pass in both directions. A normal vector with negative norm corresponds to a timelike surface. Such surfaces can only be passed in one direction, which is due to the fact that one can neither go backward in time nor side wise. For more details, please consult [1, 9].

In what follows, we look for time independent axially symmetric surfaces with a null normal vector and the surfaces fulfill the condition [1]

$$u(r, \vartheta) = \text{const.} \tag{3.110}$$

Their normal vector is given by

$$n_\alpha = \left(0, \frac{\partial u}{\partial r}, \frac{\partial u}{\partial \vartheta}, 0\right). \tag{3.111}$$

Setting the norm $n_\alpha n^\alpha = 0$ yields the equation

$$\left(r^2 - 2mr + a^2 + \frac{B}{2r}\right)\left(\frac{\partial u}{\partial r}\right)^2 + \left(\frac{\partial u}{\partial \vartheta}\right)^2 = 0, \tag{3.112}$$

which can be solved by a product ansatz $u = R(r)\Theta(\vartheta)$

$$-\left(r^2 - 2mr + a^2 + \frac{B}{2r}\right)\left(\frac{\frac{\partial R}{\partial r}}{R}\right)^2 = \left(\frac{\frac{\partial \Theta}{\partial \vartheta}}{\Theta}\right)^2. \tag{3.113}$$

Both sides of this equation depend on different variables and thus have to be constant. In analogy to [1] we will call that constant λ which then gives

$$\Theta = A e^{\sqrt{\lambda}\vartheta}. \tag{3.114}$$

This expression however is not periodic in ϑ and therefore can't describe a surface except for the case where $\lambda = 0$, for which $\Theta = \text{const.}$ The remaining equation for R is

$$\left(r^2 - 2mr + a^2 + \frac{B}{2r}\right)\left(\frac{\frac{\partial R}{\partial r}}{R}\right)^2 = 0. \tag{3.115}$$

Excluding the trivial case $\frac{\partial R}{\partial r} = 0$ we are left with the solution of

$$\left(r^2 - 2mr + a^2 + \frac{B}{2r}\right) = 0. \tag{3.116}$$

Possible physical solutions for r are given by positive real roots of the cubic polynomial

$$p(r) = r^3 - 2mr^2 + a^2 r + \frac{B}{2}. \tag{3.117}$$

Since the derivative

$$p'(r) = 3r^2 - 4mr + a^2 \tag{3.118}$$

is positive for all $r \leq 0$, from $p(0) = B/2 > 0$ and $\lim_{r \to -\infty} p(r) = -\infty$ it follows that $p(r)$ has always exactly one negative real root, which is not relevant in our case. Depending on the parameters a^2 and B there may be two more real roots, which have to be positive numbers and represent possible solutions of (3.116). This cubic function has three distinct real roots if it has a positive discriminant [17]. For $p(r)$ the parameter dependent discriminant $D(a^2, B)$ reads

$$D(a^2, B) = \frac{1}{27} \left(4 \left(4m^2 - 3a^2 \right)^3 - \left(18ma^2 - 16m^3 + \frac{27}{2} B \right)^2 \right). \tag{3.119}$$

It is easy to see that a first condition for $D(a^2, B) > 0$ is already satisfied by $a^2 < (4/3)m^2$. Rewriting the condition $D(a^2, B) > 0$, using the parametrization $a^2 = \varepsilon(4/3)m^2$, with $\varepsilon \in [0, 1]$, one obtains

$$4 \left(4(1 - \varepsilon)m^2 \right)^3 > \left[8m^3 (3\varepsilon - 2) + \frac{27}{2} B \right]^2. \tag{3.120}$$

Now, we determine the maximal parameter value B^* for which this condition is satisfied. The left hand term decreases monotonically with increasing ε, whereas the right hand term increases monotonically as long as the term in the bracket is positive. If for some ε and B the condition is met with a negative term in the bracket on the right hand side, we can choose a larger B such that this term is positive and the condition is still fulfilled. It follows that the maximum value B^* satisfying the condition (3.120) is obtained for $\varepsilon = 0$, or equivalently $a = 0$. In this case (3.120) reads

$$4 \left(4m^2 \right)^3 = \left(16m^3 \right)^2 > \left(\frac{27}{2} B - 16m^3 \right)^2, \tag{3.121}$$

which yields $B^* = (4/3)^3 m^3$. This value corresponds to the limiting case for the Schwarzschild solution including corrections proportional to $\frac{B}{2r^3}$. We conclude that also for $a > 0$, for $B > B^*$ there are no positive real roots of (3.116) and, therefore, just as in the Schwarzschild case *the modified Kerr solution shows no event horizons*. Note that there are no surfaces of infinite redshift as $g_{00} = 0$ is contained in our discussion of (3.116) because a^2 and $a^2 \cos^2 \vartheta$ have the same range of values.

The numerical studies of specific cases, like the massive object in the Sgr-A*, in the center of our galaxy, will be presented in Chap. 5 on possible experimental verification of our theory.

3.4 The pc-Reissner-Nordström Solution

In standard GR, the Reissner-Nordström solution corresponds to a charged central mass [18]. Here, we proceed in complete analogy to [1, 9]. The only difference is the additional contribution due to the dark energy-momentum tensor. If the reader is only interested in the standard GR result, he just has to set in each step the $B = bm^3$ to zero. More details can be found in [19], besides in [1, 9].

Since we consider a central, charged mass at rest, the spherical symmetry is conserved and the line element squared has the same structure as in the pc-Schwarzschild case. Furthermore, we can adopt the Einstein equation after adding the energy-momentum tensor for the electromagnetic field using the same notation for Ξ_μ as in the section on the Schwarzschild solution (see (3.33)). Hence, the Einstein equation reads

$$
\begin{aligned}
\mathscr{R}_\mu^{\ \nu} - \frac{1}{2} g_\mu^{\ \nu} \mathscr{R} &= \left(\Xi^{RN}\right)_\mu^{\ \nu} \sigma_- + \frac{8\pi\kappa}{c^2} \left(T^{em}\right)_\mu^{\ \nu} \\
&= \left(\Xi^{RN}\right)_\mu^{\ \nu} \sigma_- + \frac{8\pi\kappa}{c^2} \left(T^{em}\right)_\mu^{\ \nu} (\sigma_+ + \sigma_-) \\
&= \frac{8\pi\kappa}{c^2} \left(T^{em}\right)_\mu^{\ \nu} \sigma_+ + \left(\left(\Xi^{RN}\right)_\mu^{\ \nu} + \frac{8\pi\kappa}{c^2} \left(T^{em}\right)_\mu^{\ \nu}\right) \sigma_- \\
&= \left(\Xi^{+RN}\right)_\mu^{\ \nu} \sigma_+ + \left(\Xi^{-RN}\right)_\mu^{\ \nu} \sigma_-,
\end{aligned}
\tag{3.122}
$$

where $\left(T^{em}\right)_\mu^{\ \nu}$ is given by [1]

$$
\left(T^{em}\right)_\mu^{\ \nu} = \frac{\varepsilon^2}{2c^2 R^4}
\begin{pmatrix}
-1 & 0 & 0 & 0 \\
0 & -1 & 0 & 0 \\
0 & 0 & 1 & 0 \\
0 & 0 & 0 & 1
\end{pmatrix}
\tag{3.123}
$$

and ε depends on the charge Q in the following way

$$
\varepsilon = \frac{Q}{4\pi\varepsilon_0}.
\tag{3.124}
$$

In standard GR, instead of R the r, the real part of R appears. In (3.122) we used that $(\sigma_+ + \sigma_-) - 1$ and wrote the right hand side in terms of the zero divisor components. Finally the $\Xi_\mu^{\pm\nu\ RN}$ functions where introduced in order to obtain a symmetric expression. For both zero divisor components the equation appear now very similar and a solution can be obtained in complete analogy to the σ_- component in the Schwarzschild case!

There is, however, an important difference: The energy momentum tensor of the electro-magnetic contribution (3.123) corresponds to an anisotropic "electro-magnetic fluid", because the $T_{33}^{\nu\,em} = T_{22}^{\nu\,em} \sim p_{\vartheta}^{em}$ tangential component is the negative of the $T_{11}^{\nu\,em} \sim p_r^{em}$ radial component. Because the $T_{00}^{\nu\,em}$ is the same as the $T_{11}^{\nu\,em}$, the density is equal to the radial pressure. *The main point is that the combined electro-magnetic and dark energy tensors describe an anisotropic fluid.*

The ξ_μ^{RN} can be defined such that a relation in complete analogy to (3.30)- In the Schwarzschild case, keeps valid, i.e.

$$-\frac{1}{4}e^{-\lambda_{RN}}\xi_0^{RN} + \frac{1}{4}e^{-\lambda_{RN}}\xi_1^{RN} + \frac{\xi_2^{RN}}{R_-^2} = \Xi_0^{RN}$$

$$\frac{1}{4}e^{-\lambda_{RN}}\xi_0 - \frac{1}{4}e^{-\lambda_{RN}}\xi_1 + \frac{\xi_2^{RN}}{R_-^2} = \Xi_1^{RN}$$

$$\frac{1}{4}e^{-\lambda_{RN}}\left(\xi_0^{RN} + \xi_1^{RN}\right) = \Xi_2^{RN}. \tag{3.125}$$

We shall now solve these equations:

As in the previous chapter the σ_+ component does not differ from the usual GR field equations, which can be obtained in the same way as done in [1]. Thus we only have to solve for the σ_- component. As further above, we take into account, that the energy-momentum tensor of the electromagnetic contribution, $T_\mu^{\nu\,em}$, is real, i.e., $T_\mu^{\nu\,em} = T_\mu^{\nu\,em}(\sigma_- + \sigma_+)$. Thus the σ_+ component is the same as the σ_- component. In addition the definition of the Riemann scalar is used, namely

$$\mathscr{R} = \mathscr{R}_\mu^{\ \mu} = \mathscr{R}_0^{\ 0} + \mathscr{R}_1^{\ 1} + \mathscr{R}_2^{\ 2} + \mathscr{R}_3^{\ 3}. \tag{3.126}$$

In what follows, we restrict, as in the pc-Schwarzschild case, to the σ_- component: With (3.126) and $g_\mu^\nu = \delta_\mu^\nu$, the Einstein equations (3.122) for the σ_- component are

$$\frac{1}{2}\left(\mathscr{R}_{-0}^0 - \mathscr{R}_{-1}^1 - \mathscr{R}_{-2}^2 - \mathscr{R}_{-3}^3\right) = \Xi_0^{RN} + \tfrac{8\pi\kappa}{c^2}\left(T^{em}\right)_0^0$$

$$\frac{1}{2}\left(\mathscr{R}_{-1}^1 - \mathscr{R}_{-0}^0 - \mathscr{R}_{-2}^2 - \mathscr{R}_{-3}^3\right) = \Xi_1^{RN} + \tfrac{8\pi\kappa}{c^2}\left(T^{em}\right)_1^1$$

$$\frac{1}{2}\left(\mathscr{R}_{-2}^2 - \mathscr{R}_{-0}^0 - \mathscr{R}_{-1}^1 - \mathscr{R}_{-3}^3\right) = \Xi_2^{RN} + \tfrac{8\pi\kappa}{c^2}\left(T^{em}\right)_2^2$$

$$\frac{1}{2}\left(\mathscr{R}_{-3}^3 - \mathscr{R}_{-0}^0 - \mathscr{R}_{-1}^1 - \mathscr{R}_{-2}^2\right) = \Xi_3^{RN} + \tfrac{8\pi\kappa}{c^2}\left(T^{em}\right)_3^3. \tag{3.127}$$

Taking the difference between the third and the fourth equation in (3.127) (see Exercise 3.7) yields

$$\mathscr{R}^2_{-2} - \mathscr{R}^3_{-3} = \varXi^{RN}_2 - \varXi^{RN}_3 + \frac{8\pi\kappa}{c^2}\left(\left(T^{em}\right)^2_{\ 2} - \left(T^{em}\right)^3_{\ 3}\right) \qquad (3.128)$$

and since the spherical symmetry and (3.123) demand
$\mathscr{R}^2_{-2} - \mathscr{R}^3_{-3} = \frac{8\pi\kappa}{c^2}\left(\left(T^{em}\right)^2_{\ 2} - \left(T^{em}\right)^3_{\ 3}\right) = 0$, we obtain

$$\varXi^{RN}_2 = \varXi^{RN}_3 . \qquad (3.129)$$

The difference of the first and second equation in (3.127) leads to (see Exercise 3.7).

$$\lambda'_{RN-} + \nu'_{RN-} = R_- e^{\lambda_{RN-}}\left(\varXi^{RN}_1 - \varXi^{RN}_0\right) = \frac{R_-}{2}\left(\xi^{RN}_0 - \xi^{RN}_1\right). \qquad (3.130)$$

After differentiation we get

$$\nu''_{RN-} = -\lambda''_{RN-} + \tfrac{1}{2}\left(\xi^{RN}_0 - \xi^{RN}_1\right) + \tfrac{R_-}{2}(\xi^{RN'}_0 - \xi^{RN'}_1), \qquad (3.131)$$

which is similar to (3.47).

Adding two times the second equation of (3.127) to the last difference of the first and second equation in (3.127), finally multiplying with $e^{\lambda_{RN-}}$ yields

$$\nu''_{RN-} - \frac{\lambda'_{RN-}\nu'_{RN-}}{2} + \frac{\nu'^2_{RN-}}{2} - \frac{2\lambda'_{RN-}}{R_-} = \xi^{RN}_1 + \frac{2A}{R^4_-}e^{\lambda_{RN-}}, \qquad (3.132)$$

where we used the abbreviation (see (3.124) for the definition of ε)

$$A := +\frac{4\pi\kappa\varepsilon^2}{c^4}. \qquad (3.133)$$

Exercise 3.7 (Equations (3.130)–(3.132))

Problem. Proof (3.130)–(3.132).

Solution.
In order to verify (3.130), we subtract the second equation in (3.127) from the first one, giving

$$\left(\mathscr{R}^0_{-0} - \mathscr{R}^1_{-1}\right) = \varXi^{RN}_0 - \varXi^{RN}_1 . \qquad (3.134)$$

The contribution of the electromagnetic tensor cancels, using (3.123).
The left hand side of this equation can be rewritten:

$$\mathcal{R}^0_{-0} - \mathcal{R}^1_{-1} = g^{00}\mathcal{R}^-_{00} - g^{11}\mathcal{R}^-_{11}$$
$$= -e^{-\nu_-}\mathcal{R}^-_{00} - e^{-\lambda_-}\mathcal{R}^-_{11} \qquad (3.135)$$

Using (3.26), which expresses the diagonal components of the Ricci tensor to the ξ_μ functions for the pc-Schwarzschild case (as stated above, we can use the same steps) yields

$$-e^{-\nu_-}\left(\frac{1}{2}\right)e^{\nu_- - \lambda_-}\xi^{RN}_0 + \frac{1}{2}e^{-\lambda_-}\xi^{RN}_1 = -\frac{1}{2}e^{-\lambda_-}\left(\xi^{RN}_0 - \xi^{RN}_1\right), \quad (3.136)$$

where the index RN refers now to the Reissner-Nordstöm case and the minus index have been added in order to stress the fact that we consider the σ_- component.

Using (3.33) finally leads to

$$\Xi^{RN}_0 - \Xi^{RN}_1. \qquad (3.137)$$

This gives the right hand side of (3.134).

The left hand side of (3.134) in terms of the ν_- and λ_- functions is obtained, using (3.35), giving

$$-\left(e^{-\nu_-}\mathcal{R}^-_{00} + e^{-\lambda_-}\mathcal{R}^-_{11}\right) = -\frac{e^{-\lambda_-}}{R_-}\left(\lambda'_- + \nu'_-\right). \qquad (3.138)$$

After multiplying by $-R_-e^{\lambda_-}$ and together with (3.137) this yields (3.130).

In order to verify (3.132) we multiply the third equation in (3.127) by two and add the difference of first and second equation, which we have calculated in the previous steps of this exercise. This gives

$$\left(\mathcal{R}^2_{-2} - \mathcal{R}^0_{-0} - \mathcal{R}^1_{-1} - \mathcal{R}^3_{-3}\right) + \left(\mathcal{R}^0_{-0} - \mathcal{R}^1_{-1}\right)$$
$$= \mathcal{R}^2_{-2} - 2\mathcal{R}^1_{-1} - \mathcal{R}^3_{-3}$$
$$= g^{22}\mathcal{R}^-_{22} - 2g^{11}\mathcal{R}^-_{11} + g^{33}\mathcal{R}^-_{33}$$
$$= \frac{1}{R^2}\mathcal{R}^-_{22} - 2e^{-\lambda_-}\mathcal{R}^-_{11} + \frac{1}{R^2\sin^2\vartheta}\mathcal{R}^-_{33}. \qquad (3.139)$$

Using (3.35) gives finally

$$e^{-\lambda_-}\left[\nu''_- + \frac{(\nu'_-)^2}{2} - \frac{\lambda'_-\nu'_-}{2} - \frac{2\lambda'_-}{R}\right]. \qquad (3.140)$$

For the left hand side, using (3.26), we have

$$\frac{1}{R^2}\mathscr{R}_{22}^- - 2e^{-\lambda_-}\mathscr{R}_{11}^- + \frac{1}{R^2\sin^2\vartheta}\mathscr{R}_{33}^-$$

$$= -\frac{1}{R^2}\xi_2^{RN} + 2e^{-\lambda_-}\frac{1}{2}\xi_1^{RN} + \frac{1}{R^2\sin^2\vartheta}\xi_2^{RN}\sin^2\vartheta$$

$$= -\frac{1}{2}e^{\lambda_-}\left(\xi_0^{RN} - \xi_1^{RN}\right) + \left(2\Xi_2^{RN} + \frac{16\pi\kappa}{c^2}T_2^{em\,2}\right). \qquad (3.141)$$

Using for Ξ_2^{RN} (3.30) leads to

$$-\frac{e^{-\lambda_-}}{2}\left(\xi_0^{RN} - \xi_1^{RN}\right) + \frac{e^{-\lambda_-}}{2}\left(\xi_0^{RN} + \xi_1^{RN}\right) + \frac{16\pi\kappa}{c^2}T_2^2. \qquad (3.142)$$

Multiplying (3.140) and (3.142) with e^{λ_-} and setting them equal, gives (3.132).

After including (3.130), the sum of the equations for the index 0 and 1 in (3.127) leads to

$$\left(R_-e^{-\lambda_{RN-}}\right)' = 1 + \xi_2^{RN} - \frac{1}{4}R_-^2 e^{-\lambda_{RN-}}\left(\xi_0^{RN} - \xi_1^{RN}\right) - \frac{A}{R_-^2},$$

$$(3.143)$$

which gives after an integration

$$e^{-\lambda_{RN-}} = 1 - \frac{2M_-}{R_-} + \frac{1}{R_-}\int \xi_2^{RN}dR_-$$

$$-\frac{1}{4R_-}\int e^{-\lambda_{RN-}}R_-^2\left(\xi_0^{RN} - \xi_1^{RN}\right)dR_- + \frac{A}{R_-^2}.$$

$$(3.144)$$

With this, an equivalent equation is obtained as in (3.52) in *complete analogy*, using exactly the same steps.

$$\frac{e^{\lambda_{RN-}}}{R_-}\xi_2^{RN\prime} - \xi_1^{RN}$$

$$= \qquad (3.145)$$

$$-\frac{1}{4}R_-\left(\xi_0^{RN\prime} - \xi_1^{RN\prime}\right) + \frac{1}{2}\lambda_-^{RN\prime}R_-\left(\xi_0^{RN} - \xi_1^{RN}\right) - \frac{1}{8}R_-^2\left(\xi_0^{RN} - \xi_1^{RN}\right)^2.$$

This differential equation relates ξ_0^{RN}, ξ_1^{RN} and ξ_2^{RN} and can further be rewritten into a TOV equation (see (3.53)). This is not needed here, thus we limit ourself to (3.145).

Using (3.130) and (3.144) we can calculate the g_{00}^--component of the metric

$$g_{00}^- = -e^{\nu_{RN-}} = -e^{-\lambda_{RN-}}e^{\frac{1}{2}\int R_-\left(\xi_0^{RN} - \xi_1^{RN}\right)dR_-}. \qquad (3.146)$$

Within the ideal fluid ansatz the metric terms can be written in the form

$$\bar{g}_{11} = \left(1 - \frac{2M_-}{R_-} + \frac{2m_{de}^{RN}(R_-)}{R_-} + \frac{A}{R_-^2}\right)^{-1} \tag{3.147}$$

$$\bar{g}_{00} = -\left(1 - \frac{2M_-}{R_-} + \frac{2m_{de}^{RN}(R_-)}{R_-} + \frac{A}{R_-^2}\right) e^{f_{A-}}, \tag{3.148}$$

where, equivalent to (3.73), the function $f_{A-} = \frac{1}{2} \int R_- \left(\xi_0^{RN} - \xi_1^{RN}\right) dR_-$ has still to be determined. The function f_{A-} can be set to zero if an anisotropic fluid is assumed. This was shown in the pc-Schwarzschild case. There it also was shown that the integrand of f is proportional to $\left(\frac{p_r}{c^2} + \rho\right)$ and using the equation of state $\frac{p_r}{c^2} = -\rho$, this integrand vanishes and $e^f = e^0 = 1$. In the pc-Reissner-Nordström case, this assumption for the dark energy part can also be done.

We are now able to determine the real metric: We use the same notations as in the Schwarzschild case and additionally define $(f_A)_+ = 0$ (remember that we introduced this definition in order to have the same functional form of the metric tensor in both σ-components) and $f_A = (f_A)_+ \sigma_+ + (f_A)_- \sigma_-$. Defining

$$\Omega_{RN} = \Omega_+\sigma_+ + \Omega_-\sigma_- = \frac{B}{R_-^2}\sigma_-, \tag{3.149}$$

the complete pseudo-complex metric is given by

$$(g_{\mu\nu})$$

$$= \begin{pmatrix} -\left(1 - \frac{2\mathcal{M}}{R} + \frac{\Omega_{RN}}{R} + \frac{A}{R^2}\right) e^{f_A} & 0 & 0 & 0 \\ 0 & \left(1 - \frac{2\mathcal{M}}{R} + \frac{\Omega_{RN}}{R} + \frac{A}{R^2}\right)^{-1} & 0 & 0 \\ 0 & 0 & R^2 & 0 \\ 0 & 0 & 0 & R^2\sin^2\vartheta \end{pmatrix} \tag{3.150}$$

and the projected metric is given by

$$\begin{pmatrix} -\left(1 - \frac{2m}{r} + \frac{\Omega_{RN-}}{2r} + \frac{A}{r^2}\right) e^{\frac{f_{A-}}{2}} & 0 & 0 & 0 \\ 0 & \left(1 - \frac{2m}{r} + \frac{\Omega_{RN-}}{2r} + \frac{A}{r^2}\right)^{-1} & 0 & 0 \\ 0 & 0 & r^2 & 0 \\ 0 & 0 & 0 & r^2\sin^2\vartheta \end{pmatrix}. \tag{3.151}$$

As mentioned above, the f_{A-} can be set to zero for an anisotropic ideal fluid. For an isotropic fluid this is not the case and, therefore, it is kept in the above equations. With that we obtain the length element squared, which is

$$d\omega^2 = -\left(1 - \frac{2m}{r} + \frac{\Omega_{RN-}}{2r} + \frac{A}{r^2}\right) e^{\frac{f_{A-}}{2}} (dx^0)^2$$
$$+ \left(1 - \frac{2m}{r} + \frac{\Omega_{RN-}}{2r} + \frac{A}{r^2}\right)^{-1} (dr)^2 + r^2 \left[(d\vartheta)^2 + \sin^2\vartheta\,(d\varphi)^2\right].$$

$$(3.152)$$

Thus, both g_{00} and g_{11} do not just get a charge dependence added as in GR, but the correction term is changed as well, due to the appearance of the factor $e^{\frac{f_{A-}}{2}}$. Furthermore, all terms of g_{00} are multiplied with a charge dependent factor $e^{\frac{f_{A-}}{2}}$.

In conclusion the metric components of the Reissner-Nordström metric are not the sum of the respective components of the Schwarzschild metric and the simple GR charge term anymore. Obviously, we do predict stronger deviations to GR as a priori expected. Note, that according to (3.133) the A is *always positive*, so that the charge prevents the collapse to a singularity, as in standard GR.

References

1. R. Adler, M. Bazin, M. Schiffer, *Introduction to General Relativity*, 2nd edn. (McGraw Hill, New York, 1975)
2. G. Caspar, T. Schönenbach, P.O. Hess, M. Schäfer, W. Greiner, Pseudo-complex general relativity: Schwarzschild, Reissner-Nordstrøm and Kerr solutions. Int. J. Mod. Phys E. **21**, 1250015 (2012)
3. P. Nicolini, A. Smailagic, E. Spalluci, Noncommutative geometry inspired Schwarzschild black hole. Phys. Lett. B **632**, 547 (2006)
4. M. Visser, Gravitational vacuum polarization I: energy conditions in the Hartle-Hawking vacuum. Phys. Rev. D **54**, 5103 (1996)
5. M. Visser, Gravitational vacuum polarization II: energy conditions in the Boulware vacuum. Phys. Rev. D **54**, 5116 (1996)
6. M. Visser, Gravitational vacuum polarization IV: energy conditions in the Unruh vacuum. Phys. Rev. D **56**, 936 (1997)
7. S. Carroll, *Spacetime and Geometry* (An Introduction to General Relativity. Addison-Wesley, San Francisco, 2004)
8. P.O. Hess, W. Greiner, Pseudo-complex general relativity. Int. J. Mod. Phys. E **18**, 51 (2009)
9. C.W. Misner, K.S. Thorne, J.A. Wheeler, *Gravitation* (Freeman & Co., San Francisco, 1973)
10. T. Schönenbach, Die Kerr-Metrik in pseudokomplexer Allgemeiner Relativitatstheorie. Master's thesis, J.W. von Goethe University, Frankfurt am Main, Germany (2011)
11. B. Carter, Black hole equilibrium states. Part 1: analytic and geometric properties of Kerr solution, in *Proceedings of the 1972 Les Hoches Summer School*, ed. by C. De Witt, B.S. DeWitt (Gordon and Breach, Newark, 1973). Reprinted as a golden oldie. Gen. Relativ. Gravit. **41**, 2873 (2009)
12. J. Plebanski, A. Krasiński, *An Introduction to General Relativity and Cosmology* (Cambridge University Press, New York, 2006)
13. E. Cartan, *Geometry of Riemann Spaces* (Math Sci Press, Massachusetts, 1983)
14. C.W. Misner, The flatter regions of Newman, unti and tamburino's generalized Schwarzschild space. J. Math. Phys. **4**, 924 (1963)
15. B. O'Neill, *The Geometry of Kerr Black Holes* (A.K. Peters, wellesley, 1995)

16. N. Kamran, A. Krasinski, Editorial note: Brandon Carter, black hole equilibrium states part I: analytic and geometric properties of the Kerr solution. Gen. Relativ. Gravit. **41**, 2867 (2009)
17. M. Abramowitz, I.A. Stegun, *Handbook of Mathematical Functions* (Dover, New York, 1965)
18. G. Caspar, Die Reissner-Nordström Metric in der Pseudokomplexen Allgemainen Relativitätstheorie. Master's thesis, J.W. von Goethe University, Frankfurt am Main, Germany (2011). arXiv:1106.2653v1
19. G. Caspar, T. Schönenbach, P.O. Hess, W. Greiner, Pseudo-complex general relativity. Int. J. Mod. Phys. E **20**, 1 (2011)

Chapter 4
Pseudo-complex Robertson-Walker Metric

This chapter is mainly concerned with the topic described in [1]. There are, though, some new additions where novel solutions are presented, e.g. of an oscillating universe. Also, several corrections and improvements are presented here.

The point of partition is the so-called *Cosmological principle*. It states that the distribution of matter on the large scale can fairly be described to be homogenous. Any observer at any location in the universe should see the same physics. This principle implies also that to a good approximation one can define a global, absolute time coordinate. The fact that on the large scale no detectable deviations for the matter distribution is observed, lead to the question: *Why the matter is homogeneously distributed?* A process must happened during the early epoch of the *Big Bang* where all inhomogeneities were eliminated, a process commonly known as the *inflation*.

Let us first discuss the Robertson-Walker universe, following the steps in Chap. 12.3 of the book of Adler-Bazin-Schiffer [2]. A different signature for the metric will be used, which is in line with [3]. We shall mainly repeat these steps, for the sake of completeness, with the difference that the variables are now pseudo-complex. The reader will note that the formulation is identical to standard GR, with the difference of the appearance of additional functions due to the modified variational principle.

In order to proceed, one chooses so called *Gaussian coordinates*, in which a distinguished (absolute) time coordinate is used, thus a completely covariant treatment of the cosmological problem is abandoned [2]. This is the usual price to pay for a simplified cosmological model.

The pseudo-complex length element in Gaussian coordinates, before imposing reality, is given by

$$
\begin{aligned}
d\omega^2 &= -(dX^0)^2 + e^{G(X^0,R)}\left(dR^2 + R^2 d\vartheta^2 + R^2\sin^2\vartheta\, d\varphi^2\right) \\
&= -(dX^0)^2 + e^{G(X^0,R)} d\Sigma^2,
\end{aligned}
\tag{4.1}
$$

© Springer International Publishing Switzerland 2016   93
P.O. Hess et al., *Pseudo-Complex General Relativity*,
FIAS Interdisciplinary Science Series, DOI 10.1007/978-3-319-25061-8_4

where all coordinates are pseudo-complex. G is a function of time X^0 and the radial coordinate R. G can be written as the sum of the functions $g(X^0)$ and $f(R)$, the first one depending solely on time and the second one only on R, i.e., $G(X^0, R) = g(X^0) + f(R)$. To prove this statement, the starting point is the equivalence principle that two observers at two different points observe the same physics. The only difference may be in the scale the two observers use. Thus the ratio of the proper distance element at two different space points R_1 and R_2 must remain fixed in time:

$$\frac{e^{G(X^0, R_1)}}{e^{G(X^0, R_2)}} = \text{const in time,} \tag{4.2}$$

i.e., this ratio must be independent of X^0. Therefore one must have

$$G(X^0, R_1) = G(X^0, R_2) + F(R_1, R_2), \tag{4.3}$$

which yields for (4.2)

$$\frac{e^{G(X^0, R_2) + F(R_1, R_2)}}{e^{G(X^0, R_2)}} = e^{F(R_1, R_2)}. \tag{4.4}$$

If one chooses a fixed value for R_2, the G-function has the structure

$$G(X^0, R_1) = g(X^0) + f(R_1). \tag{4.5}$$

Next, the Christoffel symbols are determined: The equation for the geodesics is given by

$$\delta \int \left[-(\dot{X}^0)^2 + e^G \left(\dot{R}^2 + R^2\dot{\vartheta}^2 + R^2\sin^2\vartheta\,\dot{\varphi}^2 \right) \right] ds = 0, \tag{4.6}$$

with s being an affine parameter. Here, we assume that the metric is already known, having solved the Einstein equations. The equation for the geodesic is then nothing but a standard variational problem.

After variation, the following equations of motion are obtained (*a dot refers to the derivation with respect to s, the curve parameter, and a prime indicates for the function g a derivative with respect to X^0 while for f it is a derivative with respect to R*).

$$\ddot{X}^0 + \tfrac{1}{2}g'e^G \left(\dot{R}^2 + R^2\dot{\vartheta}^2 + R^2\sin^2\vartheta\,\dot{\varphi}^2 \right) = 0,$$
$$\ddot{R} + \tfrac{1}{2}f'\dot{R}^2 + g'\dot{X}^0\dot{R}$$
$$- \left(\tfrac{1}{2}f' + \tfrac{1}{R} \right) \left(R^2\dot{\vartheta}^2 + R^2\sin^2\vartheta\,\dot{\varphi}^2 \right) = 0,$$
$$\ddot{\vartheta} + 2 \left(\tfrac{1}{2}f' + \tfrac{1}{R} \right) \dot{R}\dot{\vartheta} + g'\dot{X}^0\dot{\vartheta} - \sin\vartheta\cos\vartheta\,\dot{\varphi}^2 = 0,$$

$$\ddot{\varphi} + 2\left(\tfrac{1}{2}f' + \tfrac{1}{R}\right)\dot{R}\dot{\varphi} + g'\dot{X}^0\dot{\varphi} + 2\dot{\vartheta}\dot{\varphi}\cot\vartheta = 0. \tag{4.7}$$

Comparing this with the geodesic equation

$$\ddot{X}^\mu + \left\{ \begin{matrix} \mu \\ \nu\lambda \end{matrix} \right\} \dot{X}^\nu \dot{X}^\lambda = 0, \tag{4.8}$$

yields the non-zero Christoffel symbols:

$$\left\{ \begin{matrix} 0 \\ 11 \end{matrix} \right\} = \frac{1}{2}g'e^G,$$

$$\left\{ \begin{matrix} 0 \\ 22 \end{matrix} \right\} = \frac{1}{2}g'e^G R^2,$$

$$\left\{ \begin{matrix} 0 \\ 33 \end{matrix} \right\} = \frac{1}{2}g'e^G R^2\sin^2\vartheta,$$

$$\left\{ \begin{matrix} 1 \\ 01 \end{matrix} \right\} = \frac{1}{2}g',$$

$$\left\{ \begin{matrix} 1 \\ 11 \end{matrix} \right\} = \frac{1}{2}f',$$

$$\left\{ \begin{matrix} 1 \\ 22 \end{matrix} \right\} = -R^2\left(\frac{1}{2}f' + \frac{1}{R}\right),$$

$$\left\{ \begin{matrix} 1 \\ 33 \end{matrix} \right\} = -R^2\left(\frac{1}{2}f' + \frac{1}{R}\right)\sin^2\vartheta,$$

$$\left\{ \begin{matrix} 2 \\ 02 \end{matrix} \right\} = \frac{1}{2}g' = \left\{ \begin{matrix} 3 \\ 03 \end{matrix} \right\},$$

$$\left\{ \begin{matrix} 2 \\ 12 \end{matrix} \right\} = \left(\frac{1}{2}f' + \frac{1}{R}\right) = \left\{ \begin{matrix} 3 \\ 13 \end{matrix} \right\},$$

$$\left\{ \begin{matrix} 2 \\ 33 \end{matrix} \right\} = -\sin\vartheta\cos\vartheta,$$

$$\left\{ \begin{matrix} 3 \\ 23 \end{matrix} \right\} = \cot\vartheta. \tag{4.9}$$

All others are either zero or obtained through the use of the symmetries of the Christoffel symbols.

From the line element the determinant of the metric tensor is found as

$$\ln\sqrt{-g} = \frac{3}{2}g(X^0) + \frac{3}{2}f(R) + 2\ln R + \ln|\sin|\vartheta. \tag{4.10}$$

Exercise 4.1 (Equation (4.10))

Problem. Verify (4.10).

Solution.
   The length element square is given by (4.1)

$$d\omega^2 = -\left(dX^0\right)^2 + e^{g(X^0)+f(R)}\left[(dR)^2 + R^2\,(d\vartheta)^2 + R^2\sin^2\vartheta\,(d\varphi)^2\right], \quad (4.11)$$

from which the metric matrix can be read off:

$$\begin{pmatrix} -1 & 0 & 0 & 0 \\ 0 & e^{g(X^0)+f(R)} & 0 & 0 \\ 0 & 0 & R^2 e^{g(X^0)+f(R)} & 0 \\ 0 & 0 & 0 & R^2 e^{g(X^0)+f(R)}\sin^2\vartheta. \end{pmatrix} \quad (4.12)$$

The negative of the determinant'(g) is given by

$$-g = e^{3(g(X^0)+f(R))}\,R^4\sin^2\vartheta, \quad (4.13)$$

i.e.,

$$\sqrt{-g} = e^{\frac{3}{2}(g(X^0)+f(R))}\,R^2|\sin|\vartheta \quad (4.14)$$

and its logarithm gives (4.10).

Using the Christoffel symbols, as given in (4.9), one finds

$$\left\{ \begin{matrix} \mu \\ 00 \end{matrix} \right\}_{|\mu} = 0,$$

$$\left\{ \begin{matrix} \mu \\ 11 \end{matrix} \right\}_{|\mu} = \frac{1}{2}e^G\left(g'' + g'^2\right) + \frac{1}{2}f'',$$

$$\left\{ \begin{matrix} \mu \\ 22 \end{matrix} \right\}_{|\mu} = \left[\frac{1}{2}e^G\left(g'' + g'^2\right)\right.$$
$$\left. -\left(\frac{1}{2}f'' + \frac{1}{R}f' + \frac{1}{R^2}\right)\right]R^2,$$

$$\left\{ \begin{matrix} \mu \\ 33 \end{matrix} \right\}_{|\mu} = \left[\frac{1}{2}e^G\left(g'' + g'^2\right)\right.$$
$$\left. -\left(\frac{1}{2}f'' + \frac{1}{R}f'\right)\right]R^2\sin^2\vartheta,$$
$$-\cos^2\vartheta. \quad (4.15)$$

Exercise 4.2 (Equation (4.15))

Problem. Verify (4.15).

Solution. The left hand side of the first relation in (4.15) is explicitly written, giving

$$\left\{\begin{matrix}0\\00\end{matrix}\right\}_{|0} + \left\{\begin{matrix}1\\00\end{matrix}\right\}_{|1} + \left\{\begin{matrix}2\\00\end{matrix}\right\}_{|2} + \left\{\begin{matrix}3\\00\end{matrix}\right\}_{|3}. \tag{4.16}$$

Using the list of the Christoffel symbols, as given in (4.9), one notes that all Christoffel symbols appearing in (4.16) are zero, thus the final result is zero.

Now, let us consider the second equation in (4.15): We have

$$\left\{\begin{matrix}0\\11\end{matrix}\right\}_{|0} + \left\{\begin{matrix}1\\11\end{matrix}\right\}_{|1} + \left\{\begin{matrix}2\\11\end{matrix}\right\}_{|2} + \left\{\begin{matrix}3\\11\end{matrix}\right\}_{|3}. \tag{4.17}$$

Using (4.9) we obtain with $G = g(X^0) + f(R)$

$$\frac{d}{dX^0}\left(\frac{g'}{2}e^G\right) + \frac{d}{dR}\left(\frac{f'}{2}\right) = \frac{1}{2}\left(g'' + g'^2\right)e^G + \frac{1}{2}f''. \tag{4.18}$$

For the third equation in (4.15) we have

$$\left\{\begin{matrix}0\\22\end{matrix}\right\}_{|0} + \left\{\begin{matrix}1\\22\end{matrix}\right\}_{|1} + \left\{\begin{matrix}2\\22\end{matrix}\right\}_{|2} + \left\{\begin{matrix}3\\22\end{matrix}\right\}_{|3}. \tag{4.19}$$

Using the Christoffel symbols (4.9), we obtain

$$\frac{d}{dX^0}\left(\frac{1}{2}g'e^G R^2\right) - \frac{d}{dR}\left(R^2\left[\frac{f'}{2} + \frac{1}{R}\right]\right), \tag{4.20}$$

which leads directly to the third relation in (4.15).

For the last relation in (4.15) analogous steps are applied.

The following relations are also useful

$$\left\{\begin{matrix}\mu\\0\nu\end{matrix}\right\}\left\{\begin{matrix}\nu\\0\mu\end{matrix}\right\} = \frac{3}{4}g'^2,$$

$$\left\{\begin{matrix}\mu\\1\nu\end{matrix}\right\}\left\{\begin{matrix}\nu\\1\mu\end{matrix}\right\} = \frac{1}{2}e^G g'^2 + \frac{3}{4}f'^2 + \frac{2}{R}f' + \frac{2}{R^2},$$

$$\left\{\begin{matrix}\mu\\2\nu\end{matrix}\right\}\left\{\begin{matrix}\nu\\2\mu\end{matrix}\right\} = \left[\frac{1}{2}e^G g'^2 - \frac{1}{2}f'^2 - \frac{2}{R}f'\right]$$

$$-\frac{2}{R^2} + \frac{1}{R^2}\cot^2\vartheta\Big]R^2,$$

$$\left\{\begin{matrix}\mu\\3\nu\end{matrix}\right\}\left\{\begin{matrix}\nu\\3\mu\end{matrix}\right\} = \left[\frac{1}{2}e^G g'^2 - \frac{1}{2}f'^2 - \frac{2}{R}f'\right,$$

$$-\frac{2}{R^2} + \frac{2}{R^2}\cot^2\vartheta\Big]$$

$$\times R^2\sin^2\vartheta. \tag{4.21}$$

Exercise 4.3 (Equation (4.21))

Problem. Verify (4.21)

Solution. We show the steps applied for the first relation in (4.21). For the others the steps are in complete analogy.

The left hand side of the first relation in (4.21) is given by

$$\left\{\begin{matrix}0\\0\nu\end{matrix}\right\}\left\{\begin{matrix}\nu\\00\end{matrix}\right\}$$

$$+\left\{\begin{matrix}1\\0\nu\end{matrix}\right\}\left\{\begin{matrix}\nu\\01\end{matrix}\right\} + \left\{\begin{matrix}2\\0\nu\end{matrix}\right\}\left\{\begin{matrix}\nu\\02\end{matrix}\right\}$$

$$+\left\{\begin{matrix}3\\0\nu\end{matrix}\right\}\left\{\begin{matrix}\nu\\03\end{matrix}\right\}. \tag{4.22}$$

The first term vanishes, because the Christoffel symbols are zero. What remains is

$$\left\{\begin{matrix}1\\01\end{matrix}\right\}^2 + \left\{\begin{matrix}2\\02\end{matrix}\right\}^2 + \left\{\begin{matrix}3\\03\end{matrix}\right\}^2. \tag{4.23}$$

Using (4.9) this is

$$\frac{g'^2}{4} + \frac{g'^2}{4} + \frac{g'^2}{4} = \frac{3}{4}g'^2. \tag{4.24}$$

Analogous for the other relations.

With this, the non-vanishing components of the Ricci tensor are

$$\mathcal{R}_{00} = -\frac{3}{2}g'' - \frac{3}{4}g'^2,$$

$$\mathcal{R}_{11} = -f'' - \frac{1}{R}f' + e^G\left(\frac{1}{2}g'' + \frac{3}{4}g'^2\right),$$

$$\mathcal{R}_{22} = -\left[\frac{1}{2}f'' + \frac{1}{4}f'^2 + \frac{3}{2R}f' - e^G\left(\frac{1}{2}g'' + \frac{3}{4}g'^2\right)\right]R^2,$$

$$\mathcal{R}_{33} = -\left[\frac{1}{2}f'' + \frac{1}{4}f'^2 + \frac{3}{2R}f' - e^G\left(\frac{1}{2}g'' + \frac{3}{4}g'^2\right)\right]$$
$$\times R^2\sin^2\vartheta. \tag{4.25}$$

All other components are zero.

To obtain the tensor component $\mathcal{R}_{\nu}{}^{\mu}$ the expression for the metric tensor and its inverse are needed:

$$g_{\mu\nu} = \begin{pmatrix} -1 & 0 & 0 & 0 \\ 0 & e^G & 0 & 0 \\ 0 & 0 & e^G R^2 & 0 \\ 0 & 0 & 0 & e^G R^2\sin^2\vartheta \end{pmatrix} \tag{4.26}$$

and

$$g^{\mu\nu} = \begin{pmatrix} -1 & 0 & 0 & 0 \\ 0 & e^{-G} & 0 & 0 \\ 0 & 0 & \frac{e^{-G}}{R^2} & 0 \\ 0 & 0 & 0 & \frac{e^{-G}}{R^2\sin^2\vartheta} \end{pmatrix}. \tag{4.27}$$

With this we have $(\mathcal{R}_{\mu}{}^{\nu} = g^{\nu\rho}\mathcal{R}_{\mu\rho})$

$$\mathcal{R}_0{}^0 = \frac{3}{2}g'' + \frac{3}{4}g'^2,$$

$$\mathcal{R}_1{}^1 = \left(\frac{1}{2}g'' + \frac{3}{4}g'^2\right) - e^{-G}\left(f'' + \frac{f'}{R}\right),$$

$$\mathcal{R}_2{}^2 = \mathcal{R}_3{}^3 = \left(\frac{1}{2}g'' + \frac{3}{4}g'^2\right)$$
$$- e^{-G}\left(\frac{1}{2}f'' + \frac{1}{4}f'^2 + \frac{3f'}{2R}\right). \tag{4.28}$$

The Riemann curvature scalar is

$$\mathscr{R} = \mathscr{R}_\mu{}^\mu = 3\left(g'' + g'^2\right) - 2e^{-G}\left(f'' + \frac{f'^2}{4} + \frac{2}{R}f'\right) \tag{4.29}$$

Exercise 4.4 (Equation (4.29))

Problem. Verify (4.29).

Solution. The Riemann curvature is given by

$$\begin{aligned}
\mathscr{R} &= \mathscr{R}_0{}^0 + \mathscr{R}_1{}^1 + \mathscr{R}_2{}^2 + \mathscr{R}_3{}^3 \\
&= \left(\frac{3}{2}g'' + \frac{3}{4}g'^2\right) \\
&\quad + \left(\left(\frac{1}{2}g'' + \frac{3}{4}g'^2\right) - e^{-G}\left(f'' + \frac{f'}{R}\right)\right) + 2\left[\left(\frac{1}{2}g'' + \frac{3}{4}g'^2\right)\right. \\
&\quad \left. - e^{-G}\left(\frac{1}{2}f'' + \frac{1}{4}f'^2 + \frac{3f'}{2R}\right)\right].
\end{aligned} \tag{4.30}$$

Joining terms leads to the required relation.

Denoting the energy momentum tensor by $T_\nu{}^\mu$ and exploiting the above results, the Einstein equations are

$$\frac{8\pi\kappa}{c^2}T_0{}^0 + \frac{8\pi\kappa}{c^2}T_{A0}{}^0(2\sigma_-) = \left[e^{-G}\left(f'' + \frac{f'^2}{4} + \frac{2f'}{R}\right) - \frac{3}{4}g'^2\right],$$

$$\frac{8\pi\kappa}{c^2}T_1{}^1 + \frac{8\pi\kappa}{c^2}T_{A1}{}^1(2\sigma_-) = \left[e^{-G}\left(\frac{f'^2}{4} + \frac{f'}{R}\right) - g'' - \frac{3}{4}g'^2\right],$$

$$\frac{8\pi\kappa}{c^2}T_2{}^2 + \frac{8\pi\kappa}{c^2}T_{A2}{}^2(2\sigma_-) = \left[e^{-G}\left(\frac{f''}{2} + \frac{f'}{2R}\right) - g'' - \frac{3}{4}g'^2\right],$$

$$\frac{8\pi\kappa}{c^2}T_3{}^3 + \frac{8\pi\kappa}{c^2}T_{A3}{}^3(2\sigma_-) = \left[e^{-G}\left(\frac{f''}{2} + \frac{f'}{2R}\right) - g'' - \frac{3}{4}g'^2\right],$$

$$\frac{8\pi\kappa}{c^2}T_\nu{}^\mu = 0, \quad \mu \neq \nu. \tag{4.31}$$

The 2 in front of σ_- is introduced in order to take into account the mapping to the real space, in which case the real pseudo-component of σ_-, namely $\frac{1}{2}$, is taken. On the left hand side the contribution of the dark energy due to the changed variational principle are added. Intentionally it is written in terms of an energy momentum tensor. The index Λ indicates its interpretation as a dark energy. *We stress, that within the pc-GR the appearance of the dark energy term is a consequence of the pc-description!*

Exercise 4.5 (Equation (4.31))

Problem. Verify (4.31)

Solution. Again, we show it for the first relation in (4.31):
The Einstein tensor $G_\mu^{\;\nu}$ is given by

$$G_\mu^{\;\nu} = \mathscr{R}_\mu^{\;\nu} - \frac{1}{2} g_\mu^{\;\nu} \mathscr{R}, \tag{4.32}$$

where $g_\mu^{\;\nu} = \delta_{\mu\nu}$.
For the zero-zero component this gives, using (4.28) and (4.32),

$$
\begin{aligned}
G_0^{\;0} &= \mathscr{R}_0^{\;0} - \frac{1}{2}\left(\mathscr{R}_0^{\;0} + \mathscr{R}_1^{\;1} + \mathscr{R}_2^{\;2} + \mathscr{R}_3^{\;3}\right) \\
&= \frac{1}{2}\left(\mathscr{R}_0^{\;0} - \mathscr{R}_1^{\;1} - \mathscr{R}_2^{\;2} - \mathscr{R}_3^{\;3}\right) \\
&= \frac{1}{2}\left\{\left(\frac{3}{2}g'' + \frac{3}{4}g'^2\right)\right. \\
&\quad - \left(\frac{1}{2}g'' + \frac{3}{4}g'^2\right) + e^{-G}\left(f'' + \frac{f'}{R}\right) \\
&\quad - 2\left(\frac{1}{2}g'' + \frac{3}{4}g'^2\right) \\
&\quad \left. + 2e^{-G}\left(\frac{1}{2}f'' + \frac{1}{4}f'^2 + \frac{3f'}{2R}\right)\right\}.
\end{aligned}
\tag{4.33}
$$

Joining terms, leads to the required relation.

In what follows, we will abbreviate for convenience the contribution of the dark energy by

$$\Xi_\mu = +2\frac{8\pi\kappa}{c^2} T_{\Lambda\mu}^{\;\;\mu} \tag{4.34}$$

(remember that the extra factor of 2 is due to the mapping of $2\sigma_-$ to its pseudo-real part, which is 1, and also that no summation over μ is performed in (4.34)). With this and using the relation between the energy-momentum tensor components in a fluid model to the energy density and pressure (see the pc-Schwarzschild solution in Chap. 3), the relation of the pressure and the density to the Ξ_μ-functions are

$$- 2\varepsilon_\Lambda = \frac{c^2}{8\pi\kappa}\Xi_0$$

$$2\frac{p_{\Lambda k}}{c^2} = \frac{c^2}{8\pi\kappa}\Xi_k, \qquad (4.35)$$

where p_k refers to the pressure in the variable x_k. This form allows to consider, in general, anisotropic fluid models.

4.1 Solving the Equations of Motion

Homogeneity of the matter distribution requires that

$$T_1^1 = T_2^2 = T_3^3. \qquad (4.36)$$

Because the Ξ_μ-functions are proportional to the $T_\mu{}^\mu$, the same argument can be used for the Ξ functions, i.e.,

$$\Xi_1 = \Xi_2 = \Xi_3 . \qquad (4.37)$$

The Ξ_k ($k = 1, 2, 3$) are allowed to depend on time or equivalently on the scale **a** of the radius of the universe.

Subtracting the second equation of (4.31) from the third one and exploiting the symmetry (4.36), due to which the left hand side vanishes, we arrive at the equation

$$f'' - \frac{1}{2}(f')^2 - \frac{f'}{R} = 0, \qquad (4.38)$$

i.e., *the same* equation as given in [2]. The solution is also supplied, as proposed in [2]:

$$e^f = \frac{b^2}{\left[1 - \frac{ab}{4}R^2\right]^2}, \qquad (4.39)$$

with a and b as constants.

Exercise 4.6 (Equation (4.39))

Problem. Verify (4.39).

Solution. Equation (4.38) is resolved by

$$f' = \alpha R e^{\frac{f}{2}}, \tag{4.40}$$

where α is one integration constant. That the last equation is a solution can be verified directly by inserting it into (4.38), using

$$
\begin{aligned}
f'' &= \alpha e^{\frac{f}{2}} + \frac{\alpha R}{2} f' e^{\frac{f}{2}} \\
&= \alpha e^{\frac{f}{2}} + \frac{\alpha^2 R^2}{2} e^{f} \\
&= \frac{f'}{R} + \frac{1}{2} \left(f'\right)^2 .
\end{aligned} \tag{4.41}
$$

We obtain

$$
\begin{aligned}
f'' - \tfrac{1}{2} \left(f'\right)^2 - \tfrac{f'}{R} \\
= \alpha e^{\frac{f}{2}} + \tfrac{\alpha^2 R^2}{2} e^{f} \\
- \tfrac{\alpha^2 R^2}{2} e^{f} \\
- \alpha e^{\frac{f}{2}} = 0.
\end{aligned} \tag{4.42}
$$

Redefining $|\alpha|\beta$ as $1/\mathbf{a}_0^2$, we obtain

$$e^{f} = \frac{b^2}{\left[1 + \frac{kR^2}{4\mathbf{a}_0^2}\right]^2}, \tag{4.43}$$

where $k = 0, +1$ or -1, corresponding to $\alpha\beta = 0$, negative or positive respectively.

Writing out the length element squared (4.1), we can absorb the constant β^2 into the function $e^{G(X^0)}$ and use the constant $|\alpha|\beta = \frac{1}{\mathbf{a}_0^2}$ [2].

With the redefinition of the constants α and β, as given in Exercise 4.6, the length square element takes the form

$$d\omega^2 = -(dX^0)^2 + e^{g(X^0)} \frac{1}{\left(1 + \frac{kR^2}{4a_0^2}\right)^2} d\Sigma^2. \tag{4.44}$$

The a_0 is a new constant, whose interpretation will be given later. It is exactly of the same form as in standard GR, with the difference that the coordinates are now pseudo-complex. This is in distinction to the pseudo-complex Schwarzschild metric [4], where the differences appear already in the functional form of the metric. The k acquires the values $k = 0, \pm 1$. The function $g(X^0)$ is yet undetermined. The k-values of $0, \pm 1$ can be used to model different universes. As shown in [2], the $k = 0$ corresponds to a flat universe, which is the observed curvature of our universe [5]. Therefore, the only case which we will consider here is for $k = 0$.

In what follows, the co-moving pseudo-complex coordinates [2] are used, i.e., $\dot{X}^0 = 1$ and $\dot{X}^1 = \dot{X}^2 = \dot{X}^3 = 0$, where the dot refers to the derivative with respect to the eigentime.

The energy-momentum tensor takes the form

$$\left(T_\nu^\mu\right) = \begin{pmatrix} -\varepsilon & & & \\ & \frac{p}{c^2} & & \\ & & \frac{p}{c^2} & \\ & & & \frac{p}{c^2} \end{pmatrix}, \tag{4.45}$$

for an isotropic fluid. Here, ε is the matter density and p the pressure.

The relevant functions in the length element take the form

$$e^{G(X^0,R)} = \frac{\mathbf{a}(t)^2}{\mathbf{a}_0^2 \left[1 + kR^2/(4\mathbf{a}_0^2)\right]^2},$$

$$e^{g(X^0)} = \frac{\mathbf{a}(t)^2}{\mathbf{a}_0^2},$$

$$e^{f(R)} = \frac{1}{\left[1 + kR^2/(4\mathbf{a}_0^2)\right]^2}, \tag{4.46}$$

which are obtained from the expression of the length square element, with some redefinitions. $\mathbf{a}$ is interpreted as a pseudo-complex scale for the *radius of the universe*. In the present epoch we can set $\mathbf{a}_0 = 1$.

Exercise 4.7 (Equation (4.46))

Problem. Verify (4.46).

Solution. The length element squared is given by (see (4.1))

$$d\omega^2 = -(dX^0)^2 + e^{G(X^0,R)}\left(dR^2 + R^2 d\vartheta^2 + R^2\sin^2\vartheta \, d\varphi^2\right), \quad (4.47)$$

with $G = g(X^0) + f(R)$.

The third equation in (4.46) was resolved in Exercise 4.6.

The second equation is just a definition of a scale, i.e, it redefines g.

The first equation is due to $e^G = e^{g+f}$, taking the second and last equation, for e^g and e^f respectively, of (4.46).

Let us substitute X^0 by its pseudo-real part ct. For example $\mathbf{a}(X^0)$ will be written as $\mathbf{a}(t)$. The derivative with respect to X^0 is converted into a derivation with respect to t, i.e., $\frac{d\mathbf{a}}{d(ct)} = \frac{1}{c}\frac{d\mathbf{a}}{dt} = \frac{\mathbf{a}'}{c}$.

In what follows, we will develop models for the universe, which describe how the *radius* of the universe evolves with time. The first who have done such an investigation where Friedmann [6] and Lemaitre [7]. Today we know several versions and extensions of such models, a couple we will present here.

We part from the above equations of motion (4.31). The expressions in the functions f and g and their derivatives can be re-expressed in terms of the variable R using (4.46). For example $f = -2\ln\left(1 + kR^2/(4a_0^2)\right)$ and $g = 2\ln\mathbf{a} - 2\ln a_0^2$. This and their derivatives have to be inserted into (4.31). Using the symmetry conditions of homogeneity (4.36) and (4.37) and the form of the energy-momentum tensor (4.45), the equations of motion acquire the form

$$\frac{8\pi\kappa}{c^2}\varepsilon = \Xi_0\sigma_- + \left[\frac{3k}{\mathbf{a}(t)^2} + \frac{3}{c^2}\frac{\mathbf{a}'(t)^2}{\mathbf{a}(t)^2}\right],$$
$$\frac{8\pi\kappa}{c^2}\frac{p}{c^2} = -\Xi_1\sigma_- - \left[\frac{k}{\mathbf{a}(t)^2} + \frac{\mathbf{a}'(t)^2}{c^2\mathbf{a}(t)^2} + \frac{2\mathbf{a}''(t)}{c^2\mathbf{a}(t)}\right]. \quad (4.48)$$

Exercise 4.8 (Equation (4.48))

Problem. Verify (4.48).

Solution.

From Exercise 4.6 we know that

$$f'' = \frac{f'}{R} + \frac{1}{2}\left(f'\right)^2, \quad (4.49)$$

and from the solution of $f = -2\ln\left(1 + kR^2/(4a_0^2)\right)$, we also know that

$$
\begin{aligned}
f' &= -2\frac{2kR}{4a_0^2}\frac{1}{\left(1 + \frac{kR^2}{4a_0^2}\right)} \\
&= -\frac{kR}{a_0^2}\frac{1}{\left(1 + \frac{kR^2}{4a_0^2}\right)}.
\end{aligned}
\tag{4.50}
$$

With these ingredients and that $g = 2\ln(\mathbf{a}) - 2\ln(\mathbf{a}_0)$, we have

$$
\left(f'' + \frac{(f')^2}{4} + \frac{2f'}{R}\right) = 3f'\left(\frac{f'}{4} + \frac{1}{R}\right)
$$

$$
g' = 2\frac{\mathbf{a}'}{\mathbf{a}_0}.
\tag{4.51}
$$

These are the terms which appear on the right hand side of the first equation in (4.31). Before substituting (4.51) into (4.31), the first equation in (4.51) will be simplified further. We get, using (4.50)

$$
\begin{aligned}
\frac{f'}{4} + \frac{1}{R} &= -\frac{kR}{4a_0^2}\frac{1}{\left(1 + \frac{kR^2}{4a_0^2}\right)} + \frac{1}{R} \\
&= \frac{-\frac{kR^2}{4a_0^2} + 1 + \frac{kR^2}{4a_0^2}}{R\left(1 + \frac{kR^2}{4a_0^2}\right)} \\
\xrightarrow{} \quad 3f'\left(\frac{f'}{4} + \frac{1}{R}\right) &= -\frac{3kR}{a_0^2}\frac{1}{R\left(1 + \frac{kR^2}{4a_0^2}\right)} \\
&= -\frac{3k}{\mathbf{a}^2}\frac{\mathbf{a}^2}{a_0^2}\frac{1}{\left(1 + \frac{kR^2}{4a_0^2}\right)} \\
&= -\frac{3k}{\mathbf{a}^2}e^G,
\end{aligned}
\tag{4.52}
$$

having used in the last step that $e^G = e^{g+f}$.

With all this, the right hand side in the first equation of (4.31) is given by

$$e^{-G}\left(f'' + \frac{(f')^2}{4} + \frac{2f'}{R}\right) - \frac{3}{4}(g')^2$$

$$= e^{-G}e^G\left(-\frac{3k}{\mathbf{a}^2}\right) - \frac{3}{4}4\frac{(\mathbf{a}')^2}{\mathbf{a}_0^2}$$

$$= -\left[\frac{3k}{\mathbf{a}^2} + \frac{3(\mathbf{a}')^2}{\mathbf{a}^2}\right]. \tag{4.53}$$

Using also (4.35), the left hand side of the first equation in (4.31) is

$$-\frac{8\pi\kappa}{c^2}\varepsilon + \Xi_0. \tag{4.54}$$

Combining both sides gives the first equation in (4.48). For the second equation in (4.48) the steps are analogous.

In deriving (4.54) the first Friedmann equation, given in (4.50), was used too. This equation was substituted into (4.51), which finally leads to (4.54).

Mapping to the pseudo-real part, σ_- is substituted by $\frac{1}{2}$ and from (4.48) one obtains

$$\frac{8\pi\kappa}{c^2}\varepsilon = \frac{\Xi_0}{2} + \left[\frac{3k}{\mathbf{a}_r(t)^2} + \frac{3}{c^2}\frac{\mathbf{a}_r'(t)^2}{\mathbf{a}_r(t)^2}\right],$$

$$\frac{8\pi\kappa}{c^2}\frac{p}{c^2} = -\frac{\Xi_1}{2} - \left[\frac{k}{\mathbf{a}_r(t)^2} + \frac{\mathbf{a}_r'(t)^2}{c^2\mathbf{a}_r(t)^2} + \frac{2\mathbf{a}_r''(t)}{c^2\mathbf{a}_r(t)}\right]. \tag{4.55}$$

Note that we can rewrite this equation as

$$\frac{8\pi\kappa}{c^2}(\varepsilon + \varepsilon_\Lambda) = \left[\frac{3k}{a_r(t)^2} + \frac{3}{c^2}\frac{a_r'(t)^2}{a_r(t)^2}\right],$$

$$\frac{8\pi\kappa}{c^2}\left(\frac{p}{c^2} + \frac{p_\Lambda}{c^2}\right) = -\left[\frac{k}{a_r(t)^2} + \frac{a_r'(t)^2}{c^2a_r(t)^2} + \frac{2a_r''(t)}{c^2a_r(t)}\right], \tag{4.56}$$

where (4.35) has been used.

After projection, the pseudo-complex approach in the limit taken here is thus equivalent to the classical case with an additional energy-momentum tensor with energy density ε_Λ and pressure p_Λ, which can be interpreted as a dark energy.

In the next couple of pages we search for an approximate relation between $\frac{p_\Lambda}{c^2}$ and ε_Λ, in order to estimate later an improved solution: The Hubble constant is defined as

$$\frac{\mathbf{a}_r'}{\mathbf{a}_r} = H \text{ with } H' \ll 1, \tag{4.57}$$

where the prime refers to the derivative with respect to time. Because $\mathbf{a} = \mathbf{a}_r + l\mathbf{a}_I$, $\mathbf{a}_I$ being the pseudo-imaginary component of $\mathbf{a}$, and l is the length parameter of the theory, which is extremely small (see (I)), one also assumes that $\mathbf{a} \approx \mathbf{a}_r$ and, because $\mathbf{a}_I = \frac{1}{2}(\mathbf{a}_+ - \mathbf{a}_-)$, we can set $\mathbf{a}_\pm \approx \mathbf{a}_r$. Using this and assuming $H' \ll 1$ we can approximately write, assuming a nearly constant $H = \frac{\mathbf{a}_r'}{\mathbf{a}_r}$,

$$\frac{\mathbf{a}_r''}{\mathbf{a}_r} = \frac{\mathbf{a}_r''}{\mathbf{a}_r'}\frac{\mathbf{a}_r'}{\mathbf{a}_r} = (\ln\mathbf{a}_r')'\frac{\mathbf{a}_r'}{\mathbf{a}_r} = [\ln(H\mathbf{a}_r)]'\frac{\mathbf{a}_r'}{\mathbf{a}_r}$$

$$= [\ln H + \ln\mathbf{a}_r]'\frac{\mathbf{a}_r'}{\mathbf{a}_r}$$

$$\approx \left(\frac{\mathbf{a}_r'}{\mathbf{a}_r}\right)^2 = H^2. \tag{4.58}$$

With that, utilizing (4.56) and $k = 0$ and neglecting the contributions of ε and p/c^2 (for dust, $p = 0$ and the density is usually small), we can write the Ξ_0 and Ξ_1 approximately as

$$-\frac{8\pi\kappa}{c^2}\varepsilon_\Lambda = -\frac{3}{c^2}H^2$$

$$\frac{8\pi\kappa}{c^2}\frac{p_\Lambda}{c^2} \approx -\frac{3}{c^2}H^2$$

$$\rightarrow \frac{p_\Lambda}{c^2} \approx -\varepsilon_\Lambda. \tag{4.59}$$

Further below we will see that this exactly corresponds to the case of a cosmological constant not changing with the redshift. Knowing that H is changing in time, implies that there *must* be a dependence on the radius of the universe, i.e., the redshift z. For that we have to solve the equation of motion exactly. As we will see further below, it is not easy to get a relation between p_Λ and ε_Λ but rather the use of a parametrization is appropriate.

Let us now continue to solve the pc-RW model:

Let us take first the first equation of (4.56) plus three times the second equation of (4.56) and then let us take the first equation of (4.48) plus the second one, dividing by a global 2 and using the projection $\sigma_- \rightarrow \frac{1}{2}$. This leads to two equations

$$\frac{4\pi\kappa}{c^2}\left(\varepsilon + \frac{3p}{c^2}\right) = -\frac{4\pi\kappa}{c^2}\left(\varepsilon_\Lambda + \frac{3p_\Lambda}{c^2}\right) - \frac{3\mathbf{a}_r''}{c^2\mathbf{a}_r} \tag{4.60}$$

$$\frac{4\pi\kappa}{c^2}\left(\varepsilon + \frac{p}{c^2}\right) = -\frac{4\pi\kappa}{c^2}\left(\varepsilon_\Lambda + \frac{p_\Lambda}{c^2}\right) + \frac{k}{\mathbf{a}_r^2} + \frac{\mathbf{a}_r'^2 - \mathbf{a}_r\mathbf{a}_r''}{c^2\mathbf{a}_r^2}. \tag{4.61}$$

The following equation is also useful

$$\frac{\mathbf{a}_r \mathbf{a}_r'' - \mathbf{a}_r'^2}{c^2 \mathbf{a}_r^2} = \frac{d}{dt}\left(\frac{\mathbf{a}_r'}{c^2 \mathbf{a}_r}\right), \tag{4.62}$$

which can be verified directly. With this, we arrive for the second equation in (4.61) at

$$\frac{d}{dt}\left(\frac{\mathbf{a}_r'}{c^2 \mathbf{a}_r}\right) = \frac{k}{\mathbf{a}_r^2} - \frac{4\pi\kappa}{c^2}\left[\left(\varepsilon + \frac{p}{c^2}\right) + \left(\varepsilon_\Lambda + \frac{p_\Lambda}{c^2}\right)\right]. \tag{4.63}$$

Differentiation of the first equation in (4.48) with respect to time gives

$$\frac{8\pi\kappa}{c^2}\frac{d(\varepsilon + \varepsilon_\Lambda)}{dt} = -\frac{6k}{\mathbf{a}_r^3}\mathbf{a}_r' + \frac{6\mathbf{a}_r'}{\mathbf{a}_r}\frac{d}{dt}\left(\frac{1}{c^2}\frac{\mathbf{a}_r'}{\mathbf{a}_r}\right). \tag{4.64}$$

Substituting (4.63) into (4.64) and multiplying the result by $\frac{c^2}{8\pi\kappa}\mathbf{a}_r^3$, yields

$$\mathbf{a}_r^3\frac{d(\varepsilon + \varepsilon_\Lambda)}{dt} = -3\mathbf{a}_r^2\mathbf{a}_r'\left[(\varepsilon + \frac{p}{c^2}) + (\varepsilon_\Lambda + \frac{p_\Lambda}{c^2})\right]. \tag{4.65}$$

Note that $3\mathbf{a}_r^2\mathbf{a}_r' = \frac{d\mathbf{a}_r^3}{dt}$. Shifting the last term of this equation to the left hand side leads to

$$\frac{d}{dt}\left(\varepsilon\mathbf{a}_r^3\right) + \frac{p}{c^2}\frac{d\mathbf{a}_r^3}{dt} = -\frac{d}{dt}\left(\varepsilon_\Lambda\mathbf{a}_r^3\right) - \frac{p_\Lambda}{c^2}\frac{d\mathbf{a}_r^3}{dt}. \tag{4.66}$$

Exercise 4.9 (Equation (4.66))

Problem. Verify (4.66).

Solution. Shifting the last term on the right hand side in (4.65) to the left hand side of the equation gives

$$\mathbf{a}_r^3\frac{d\varepsilon}{dt} + 3\mathbf{a}_r^2\mathbf{a}_r'\left(\varepsilon + \frac{p}{c^2}\right) = \left[\mathbf{a}_r^3\frac{d\varepsilon}{dt} + \frac{d\mathbf{a}_r^3}{dt}\varepsilon\right] + \frac{dR_r^3}{dt}\left(\frac{p}{c^2}\right)$$

$$= \frac{d}{dt}\left(\varepsilon\mathbf{a}_r^3\right) + \frac{dR_r^3}{dt}\left(\frac{p}{c^2}\right). \tag{4.67}$$

We have used $\frac{d\mathbf{a}_r^3}{dt} = 3\mathbf{a}_r^2\mathbf{a}_r'$.

For the terms remaining on the right hand side of (4.64) we apply the same rule and obtain (4.66).

Identifying the mass within a given volume of the universe by $M = \varepsilon V$, with $V = \mathbf{a}_r^3$ as a given volume, the last equation can be written as

$$\frac{dM}{dt} + \frac{p}{c^2} \frac{dV}{dt} = -\frac{d\varepsilon_\Lambda V}{dt} - \frac{p_\Lambda}{c^2} \frac{dV}{dt}. \qquad (4.68)$$

This is a local energy balance! In order to maintain local energy conservation separately for the baryonic mass (left hand side of the equation) and for the dark energy (right hand side of the equation), we have to require that the right and left hand side have to be zero independently. This reflects the assumption that dark energy and mass do not exchange energy. This leaves us with the condition

$$-\frac{d\varepsilon_\Lambda}{dt} = \frac{d(\ln \mathbf{a}_r^3)}{dt} \left(\frac{p_\Lambda}{c^2} + \varepsilon_\Lambda \right). \qquad (4.69)$$

Any solution for ε_Λ and p_Λ has to fulfill this differential equation. Using $\frac{p_\Lambda}{c^2} = -\varepsilon_\Lambda$ leads to $\frac{d\varepsilon_\Lambda}{dt} = 0$, or $\varepsilon_\Lambda = \Lambda = $ const. I.e., for this case we recover the model with a cosmological constant not changing with time. This equation is not sufficient to solve for ε_Λ and $\frac{p_\Lambda}{c^2}$; in fact one condition is missing.

The equation in (4.61) has the usual interpretation when $\varepsilon_\Lambda = p_\Lambda = 0$. Then the left hand side is the sum of two positive quantities, the density and the pressure. The right hand side of (4.61) is proportional to the acceleration $\mathbf{a}_r''$ of the radius of the universe, $\mathbf{a}_r$, multiplied by (-1). This equation tells us that the acceleration of $\mathbf{a}_r$ has to be negative, i.e., we get a decelerated universe. *In contrast, in the pseudo-complex description there is an additional term in p_Λ and ε_Λ present, which might be positive.* Transferring it to the left hand side may give in total a negative function in time, i.e., *depending of the functional form of ε_Λ and $\frac{p_\Lambda}{c^2}$ in time, an accelerated phase may be reproduced or not.* Even an oscillating universe can be obtained, choosing an adequate equation of state, as will be shown further below.

Let us first verify whether we can get also acceleration, i.e., that $\mathbf{a}_r'' > 0$ in the pseudo-complex version of GR:

Using the left hand side of (4.66) (the right hand side is set to zero as argued below (4.68)), we obtain, after multiplying with dt,

$$\mathbf{a}_r^3 d\varepsilon + 3\mathbf{a}_r^2 \varepsilon d\mathbf{a}_r + \frac{p}{c^2} 3\mathbf{a}_r^2 d\mathbf{a}_r = 0. \qquad (4.70)$$

Dividing by $3\mathbf{a}_r^3$ we obtain

$$\frac{d\varepsilon}{3} + \left(\varepsilon + \frac{p}{c^2} \right) \frac{d\mathbf{a}_r}{\mathbf{a}_r} = 0. \qquad (4.71)$$

Finally, dividing by $(\varepsilon + \frac{p}{c^2})$ yields

$$\frac{d\varepsilon}{3\left(\varepsilon + \frac{p}{c^2}\right)} + \frac{d\mathbf{a}_r}{\mathbf{a}_r} = 0. \tag{4.72}$$

Now we have to make an assumption on the *equation of state*! **This is a delicate part** and the results can change, depending on which equation of state we take. The equation of state may also depend on different time epochs (as we will use for the solution of an oscillating universe). The basic assumptions are that i) the distribution of the mass in the universe can be treated as an ideal gas, dust or radiation, the mass being equally distributed (because locally there are mass concentrations, as galaxies and galaxy clusters, the homogeneous assumption is only approximately true). The equation of state is

$$\frac{p}{c^2} = \alpha\varepsilon, \tag{4.73}$$

where ε is the *energy density* and α is zero for a model with dust, $\frac{2}{3}$ for a classical ideal gas and $\frac{1}{3}$ for a relativistic ideal gas (radiation). For dust, there is no pressure, i.e. $\alpha = 0$. For the cases of a classical ideal gas and for radiation (ultra-relativistic gas) the thermodynamical equation of states have the $\alpha = \frac{2}{3}$ and $\frac{1}{3}$ respectively [8]
With this, (4.72) can be solved with the solution

$$\varepsilon = \varepsilon_0 \mathbf{a}_r^{-3(1+\alpha)}, \tag{4.74}$$

where the ε_0 is a pseudo-complex integration constant, referring to its value at the present epoch.

When this result is substituted into the first equation of (4.61), solving for $\mathbf{a}_r''$, one finds

$$\mathbf{a}_r'' = -\frac{4\pi\kappa}{3}\left(\frac{3p_\Lambda}{c^2} + \varepsilon_\Lambda\right)\mathbf{a}_r - \frac{4\pi\kappa}{3}(1+3\alpha)\varepsilon_0 \mathbf{a}_r^{-(2+3\alpha)}. \tag{4.75}$$

We will also need the relation

$$\frac{d\ln V}{dt} = \frac{1}{V}\frac{dV}{dt} = \frac{1}{\mathbf{a}_r^3}\frac{d}{dt}\mathbf{a}_r^3 = \frac{3\mathbf{a}_r'}{\mathbf{a}_r}$$

$$= \frac{d\ln \mathbf{a}_r^3}{dt}. \tag{4.76}$$

Now we remember our former result that $\frac{p_\Lambda}{c^2}$ has to be *approximately* equal to $-\varepsilon_\Lambda$ (see (4.59)). Due to this we can assume that the following relation also holds approximately:

$$\frac{p_\Lambda}{c^2} = -\beta\varepsilon_\Lambda, \tag{4.77}$$

where β is an additional parameter of the theory, describing the deviation from a constant Hubble parameter H. In principle, one can also use a power expansion of $\frac{p_\Lambda}{c^2}$ in terms of ε_Λ, or a power expansion in $\mathbf{a}_r$, which would only introduce more parameters. The β will later be related to observable quantities, like the Hubble constant. Equation (4.77) gives us the missing condition, with the prize of having to introduce an additional parameter.

Using (4.69), we obtain for the differential equation for ε_Λ

$$\frac{d\varepsilon_\Lambda}{dt} = (\beta - 1)\frac{d(\ln \mathbf{a}_r^3)}{dt}\varepsilon_\Lambda$$
$$= \frac{d(\ln \mathbf{a}_r^{3(\beta-1)})}{dt}\varepsilon_\Lambda, \tag{4.78}$$

with the solution

$$\varepsilon_\Lambda = \Lambda \mathbf{a}_r^{3(\beta-1)}. \tag{4.79}$$

Equation (4.79) leaves us with the two, yet undetermined, parameters Λ and β. There are several scenarios:

(i) $\beta = 1$: Then $\varepsilon_\Lambda = \Lambda$ is constant.
(ii) $\beta \neq 0$: This will lead (see further below) to decelerated and accelerated systems, depending on the value of β. Also the acceleration as a function of the radius of the universe (which can be correlated to time of evolution) depends on β and Λ.

Using (4.77) and (4.79) gives the final form of the equation of motion for the radius of the universe (see (4.75))

$$\mathbf{a}_r'' = \frac{4\pi\kappa}{3}(3\beta - 1)\Lambda\mathbf{a}_r^{3(\beta-1)+1}$$
$$- \frac{4\pi\kappa}{3}(1 + 3\alpha)\varepsilon_0\mathbf{a}_r^{-3(1+\alpha)+1}. \tag{4.80}$$

Exercise 4.10 (Equation (4.80))

Problem. Verify (4.80).

Solution.
We start from (4.75) and substitute ε_Λ using (4.79). This gives

$$\mathbf{a}_r'' = -\frac{4\pi\kappa}{3}\left(3\frac{p_\Lambda}{c^2} + \varepsilon_\Lambda\right)\mathbf{a}_r^{(3\beta-2)} - \frac{4\pi\kappa}{3}(1 + 3\alpha)\varepsilon_0\mathbf{a}_r^{-(2+3\alpha)}$$
$$= \frac{4\pi\kappa}{3}(3\beta - 1)\varepsilon_\Lambda\mathbf{a}_r^{(3\beta-2)} - \frac{4\pi\kappa}{3}(1 + 3\alpha)\varepsilon_0\mathbf{a}_r^{-(2+3\alpha)}. \tag{4.81}$$

Substituting ε_Λ, using the solution (4.79) and rewriting the exponent $-(2 + 3\alpha)$ of $\mathbf{a}_r$ in the second term as $[-(3 + 3\alpha) + 1]$, leads to (4.80).

4.2 Some Consequences

When $\beta = 1$ (cosmological constant), the sign of the first term in (4.80) is positive and contributes to the acceleration of the universe. The acceleration increases with the radius of the universe. For a general β, the acceleration is positive, as long as $\beta > \frac{1}{3}$ and it is negative (deceleration) for $\beta < \frac{1}{3}$. For $\beta = \frac{1}{3}$ no additional acceleration nor deceleration takes place. The last term in (4.80) is always negative. How the accelerating term behaves as a function in $\mathbf{a}_r$ is also determined by β. If the exponent of $\mathbf{a}_r$ is positive, the acceleration increases with $\mathbf{a}_r$, for $\beta > \frac{2}{3}$, while it decreases with $\mathbf{a}_r$ for $\beta < \frac{2}{3}$.

Let us now discuss several particular values of β. The Λ is taken as positive [5] and the equation of state is $\frac{p_\Lambda}{c^2} = -\varepsilon_\Lambda$.

We can now rewrite (4.80) by, dividing by $\left(\frac{4\pi\kappa}{3}\right)\varepsilon_0$, and have, i.e., $\alpha = 0$:

$$\widetilde{\mathbf{a}}_r'' = \frac{\mathbf{a}_r''}{\left(\frac{4\pi\kappa}{3}\right)\varepsilon_0}$$
$$= \Lambda(3\beta - 1)\,\mathbf{a}_r^{3(\beta-1)+1} - (1 + 3\alpha)\,\mathbf{a}_r^{-2}. \qquad (4.82)$$

In the following, we discuss the case of a dust dominated universe, i.e., $\alpha = 0$. For the case of a relativistic ideal gas $\alpha = \frac{1}{3}$, the results show the same characteristics. We take arbitrarily different values of β, which are chosen such that we have as a special case the cosmological *constant* Λ and further possibilities where the dark energy density varies over time. New solutions will arise, which are not necessarily present in nature, i.e., β might have a different intermediate value as those presented in the following examples. The discussion is mainly of conceptual nature. With this we get for

(a) $\beta = 1$: ($\varepsilon_\Lambda = \Lambda$, i.e., the case of a constant dark energy density)

$$\widetilde{\mathbf{a}}_r'' = 2\Lambda\mathbf{a}_r - \mathbf{a}_r^{-2}. \qquad (4.83)$$

The universe is accelerated by the first contribution and decelerated by the second one. For small $\mathbf{a}_r$ the second term dominates and the universe is decelerated. For large $\mathbf{a}_r$ the first term starts to dominate and the universe is from then on accelerated. The turning point ($\mathbf{a}_r'' = 0$) is at

$$\mathbf{a}_r = \frac{1}{(2\Lambda)^{\frac{1}{3}}} \approx 0.79/\Lambda^{\frac{1}{3}}. \qquad (4.84)$$

If we set the scale of the radius of the present epoch to $\mathbf{a}_0 = 1$, the result implies that for $\Lambda = 1$ acceleration sets in after the universe passed 80 % of its radius. This case corresponds to a constant cosmological function Λ.

(b) $\beta = \frac{4}{3}$:

Using $\alpha = 0$ in (4.82) leads to

$$\widetilde{\mathbf{a}}_r'' = 3\Lambda \mathbf{a}_r^2 - \mathbf{a}_r^{-2}. \tag{4.85}$$

In this case, the acceleration increases with the second power in $\mathbf{a}_r$, stronger than only with a cosmological constant. The ε_Λ function (4.79) is then given by $\Lambda \mathbf{a}_r$, i.e., the dark energy density, increases with the radius of the universe. As a consequence, repulsion would increase indefinitely, finally breaking all structures, like galaxies, planetary systems and even atoms, apart. This is known in the literature as the so-called *big rip-off*.

The break-even point ($\mathbf{a}_r'' = 0$) is reached for

$$\mathbf{a}_r = 1/(3\Lambda)^{\frac{1}{4}} \approx \frac{0.76}{\Lambda^{\frac{1}{4}}}. \tag{4.86}$$

For $\Lambda = 1$ the break-even point is reached when the universe is $1/3$ of its present radius, thus, sightly earlier than in case a).

(c) $\beta = \frac{1}{2}$: Remember that $\alpha = 0$ (dust dominated universe)! Then, from (4.82) we get

$$\widetilde{\mathbf{a}}_r'' = \frac{1}{2}\Lambda \mathbf{a}_r^{-\frac{1}{2}} - \mathbf{a}_r^{-2}. \tag{4.87}$$

In this situation, the dark energy density behaves as (use (4.79)) $\varepsilon_\Lambda = \frac{\Lambda}{\mathbf{a}_r^{3/2}}$, i.e., the density of the dark energy decreases with the radius scale (time) of the universe.

This is really a new solution! The accelerating and the decelerating parts are decreasing with the size of the universe, *but at a different rate*. For small $\mathbf{a}_r$, the second term dominates and the universe is decelerated, while for sufficient large $\mathbf{a}_r$ the first, accelerating, term dominates and the universe is accelerated! The break-even point ($\mathbf{a}_r''$) is at

$$\mathbf{a}_r \approx 2^{\frac{2}{3}}/\Lambda^{\frac{2}{3}}, \tag{4.88}$$

i.e., for $\Lambda = 1$ the universe at this point will be at about $2^{\frac{2}{3}} \approx 1.59$ times of its present radius. For $\Lambda = 3$ the break-even point is at 76 % of the radius of the universe, which is plotted in Fig. 4.1. The universe starts accelerating after having reached $\mathbf{a}_r = 0.76$ (units in $\mathbf{a}_{r0}$). However, having reached the radius $\mathbf{a}_r \approx 1.9$, i.e., nearly twice the actual radius of the universe, it reaches a maximum and after that the acceleration is *decreasing*, reaching asymptotically zero. The position of the maximum is well appreciated in Fig. 4.1 by the horizontal line. Also the slow decrease with larger

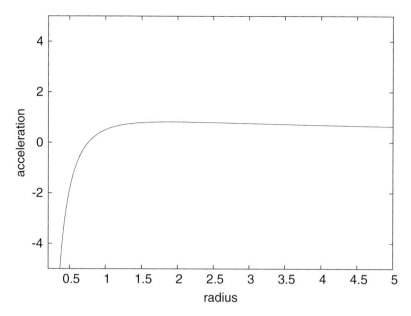

Fig. 4.1 Dependence of the scaled acceleration as a function of $\mathbf{a}_r$ (*solid line*), the radius of the universe, for $\beta = \frac{1}{2}$. In this figure $\Lambda = 3$. The maximum of this function can be deduced from (4.87), giving $\mathbf{a}_{\max} = (8/\Lambda)^{\frac{2}{3}} = 1.923$. The maximum can be barely seen in the figure due to the extreme slow decrease of the function

radii can be seen. This universe will never collapse but reach an asymptotically non-accelerating state.

The maximum for the acceleration (4.87) is obtained, setting the derivative of (4.87) to zero. This gives the equation

$$-\frac{\Lambda}{4\mathbf{a}_r^{\frac{3}{2}}} + \frac{2}{\mathbf{a}_r^3} = 0. \tag{4.89}$$

Resolving this equation leads to

$$\mathbf{a}_{\max} = \left(\frac{8}{\Lambda}\right)^{\frac{2}{3}}. \tag{4.90}$$

(d) $\beta = \frac{2}{3}$: Then, from (4.82) we get

$$\widetilde{\mathbf{a}}_r'' = \Lambda - \mathbf{a}_r^{-2}. \tag{4.91}$$

This is also a new solution. The break-even point is now at

$$\mathbf{a}_r \approx 1/\sqrt{\Lambda}. \tag{4.92}$$

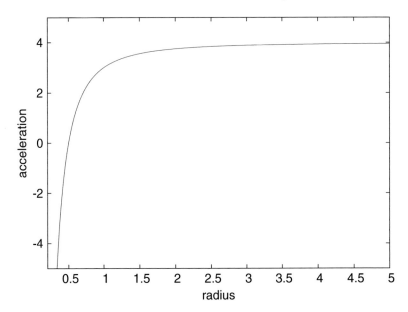

Fig. 4.2 Dependence of the scaled acceleration as a function in $\mathbf{a}_r$, the radius of the universe, for $\beta = \frac{2}{3}$. In this figure, the $\Lambda = 4$

This solution is also special in the sense that the asymptotic acceleration of the universe *is constant* ($\mathbf{a}_r'' = \Lambda$). Using (4.79) leads to the dependence $\varepsilon_\Lambda = \Lambda/\mathbf{a}_r$ of the dark energy density on the radius of the universe, i.e., it also decreases with the radius scale (time) of the universe. For the case just studied, the acceleration of the universe versus its radius is depicted in (Fig. 4.2).

In all cases Λ can be fitted to the observation at which stage the net acceleration did set in, overcoming the deceleration term in (4.82). The new part here is that other solutions exist than the standard ones:

(i) There is the possibility of a constant asymptotic acceleration.
(ii) In another solution, the expansion of the universe, after its decelerating period, gets accelerated. The accelerations reach a maximum and vanish asymptotically. In this case the universe approaches, for large times, an ever expanding, non-accelerating phase.
(iii) In all cases, the universe is first decelerated and after a so-called *break-even point* it starts to accelerate.
(iv) Of course, all standard solutions are obtained (cosmological constant and rip-off).

In order to calculate numerically observable consequences, we have to know the exact form of ε_Λ, which we were unable to deduce from first principles. One possibility is to use the calculated distribution of dark energy, as for example done in [9]. An alternative is to use the parametrization given in (4.77). This implies the use of an additional parameter (β) and is equivalent to known considerations in the literature [5]. The acceleration in each solution is a consequence of the Ξ_k functions. As discussed above, **they represent contributions to the energy-momentum tensor, which can be interpreted as dark energy. This** *dark energy* **stems from modified equations of motion for the universe due to the pseudo-complex formulation of the theory. In the model considered, this dark energy is in general not a constant but may vary in time, i.e., with the radius of the universe.** The last statement is important for an oscillating universe, as an additional solution.

Dependence of β:

The question is now: Can we say something about the dependence of β on time, or -equivalently- on the radius scale $\mathbf{a}_r$? This will be now investigated:

For that, we start from (4.58), without the approximation in the last line. Applying the derivative and with $H = \frac{\mathbf{a}_r'}{\mathbf{a}_r}$, we get

$$
\begin{aligned}
\frac{\mathbf{a}_r''}{\mathbf{a}_r} &= \left[\frac{H'}{H} + \frac{\mathbf{a}_r'}{\mathbf{a}_r} \right] \frac{\mathbf{a}_r'}{\mathbf{a}_r} \\
&= \left[\frac{H'}{H} + H \right] H = H' + H^2.
\end{aligned}
\tag{4.93}
$$

With this and (4.56) we obtain for $\frac{p_\Lambda}{c^2}$ and ε_Λ $(k = 0)$

$$
\begin{aligned}
-\frac{8\pi\kappa}{c^2} \frac{p_\Lambda}{c^2} &= \frac{1}{c^2} H^2 + \frac{2}{c^2} \left(H' + H^2 \right) + \frac{8\pi\kappa}{c^2} \left(\frac{p}{c^2} \right) \\
&= \frac{3}{c^2} H^2 + \frac{2}{c^2} H' + \frac{8\pi\kappa}{c^2} \left(\frac{p}{c^2} \right), \\
\frac{8\pi\kappa}{c^2} \varepsilon_\Lambda &= \frac{3}{c^2} H^2 - \frac{8\pi\kappa}{c^2} \varepsilon.
\end{aligned}
\tag{4.94}
$$

Exercise 4.11 (Equations (4.93) and (4.94))

Problem. Verify (4.93) and (4.94).

Solution.
We start from (4.58) and use some trivial manipulations:

$$
\begin{aligned}
\frac{\mathbf{a}_r''}{\mathbf{a}_r} &= [\ln H + \ln \mathbf{a}_r]' \frac{\mathbf{a}_r'}{\mathbf{a}_r} \\
&= \left[\frac{H'}{H} + \frac{\mathbf{a}_r'}{\mathbf{a}_r} \right] \frac{\mathbf{a}_r'}{\mathbf{a}_r},
\end{aligned}
\tag{4.95}
$$

where we have used the properties of the ln-function under derivations.

Using the definition of the Hubble function, i.e., $H = \frac{\mathbf{a}_r'}{\mathbf{a}_r}$, we obtain

$$\left[\frac{H'}{H} + H\right]H = H' + H^2. \tag{4.96}$$

For reproducing (4.94), we part from the second equation in (4.55), setting $k = 0$ (which is the condition for a flat universe), resolving for $\frac{p_\Lambda}{c^2}$. We obtain

$$-\frac{8\pi\kappa}{c^2}\frac{p_\Lambda}{c^2} = \frac{8\pi\kappa}{c^2}\frac{p}{c^2} + \left[\frac{\mathbf{a}_r'(t)^2}{c^2\mathbf{a}_r(t)^2} + \frac{2\mathbf{a}_r''(t)}{c^2\mathbf{a}_r(t)}\right] \tag{4.97}$$

Finally, substituting $\frac{\mathbf{a}_r'}{\mathbf{a}_r}$ by the Hubble function H and $\frac{\mathbf{a}_r''}{\mathbf{a}_r}$ by the above equation in terms of H and its derivative, leads to the first equation in (4.94).

The same steps for the first equation in (4.55), for ε leads to the second equation in (4.94).

Parting from (4.94), we get (see Exercise 4.11).

$$-3\frac{8\pi\kappa}{c^2}\frac{p_\Lambda}{c^2} = \frac{1}{c^2}H^2 + \frac{2}{c^2}\left(H' + H^2\right)$$

$$= \frac{3}{c^2}H^2 + \frac{2}{c^2}H',$$

$$3\frac{8\pi\kappa}{c^2}\varepsilon_\Lambda = \frac{3}{c^2}H^2. \tag{4.98}$$

The ratio of $\frac{p_\Lambda}{c^2}$ with ε_Λ, which is β, gives

$$\frac{\frac{p}{c^2}}{-\varepsilon_\Lambda} = \beta = 1 + \frac{2}{3}\frac{H'}{H^2}. \tag{4.99}$$

The consequence of this equation is that β is always different from one and larger than 1 when $H' > 0$. The relation to the w parameter, as defined in literature by $\frac{p_\Lambda}{c^2} = w\varepsilon_\Lambda$ [5], is obtained using (4.99) and $\frac{p_\Lambda}{c^2} = -\varepsilon_\Lambda$, i.e. $w = -1$ and thus $\beta = -w = +1$ (see also below (4.115))!

In conclusion, a measurement of the change of the Hubble constant with time will lead to a determination of the parameter β as a function of time. Though, the last considerations clarify the role of β, we are suffering still by the problem that we have to know the solution of $H = \frac{\mathbf{a}_r'}{\mathbf{a}_r}$. This can be done up to now only through the experimental measurement of the Hubble parameter H.

A model including dust and radiation, k = 0:

Up to now, we did only consider one density component (dust or radiation) and the pseudo-complex contribution. Realistic models involve both components, as can be seen in [5]. Expressing the ratio of the radii $\mathbf{a}_{r0}$ and $\mathbf{a}_r$, the present radius of the universe and the one at a redshift z respectively, in terms of the redshift z itself, gives [2, 5]

$$\frac{\mathbf{a}_{r0}}{\mathbf{a}_r} = (1 + z). \tag{4.100}$$

We obtain for the square of the ratio of the velocity and of the scale of the universe [5]

$$\left(\frac{\mathbf{a}_r'}{\mathbf{a}_r}\right)^2 = H^2 = H_0^2 \left\{ \Omega_d (1 + z)^3 + \Omega_r (1 + z)^4 + \Omega_\Lambda f(z) \right\}, \tag{4.101}$$

where the index d refers to the dust part and the index r to the radiation part. We do not include the contribution due to $k \neq 0$, because we consider a flat universe. The factor H_0^2 is the square of the present Hubble constant.

Exercise 4.12 (Equation (4.101))

Problem. Verify (4.101).

Solution. In this exercise we will treat two contributions only, namely from a mass with the equation of state $p = \alpha \varepsilon$ plus from the dark energy. When more contributions, like radiation, are present, one has just to repeat the "mass" term with different values of α and sum up. For example. when dust ($\alpha = 0$) and radiation ($\alpha = \frac{1}{3}$) are considered, one has to sum the expression for dust and radiation.

Let us start from the first equation in (4.61) and substitute $\frac{p_\Lambda}{c^2}$ by $-\beta \varepsilon_\Lambda = -\beta \Lambda \mathbf{a}_r^{2(\beta-1)}$ (see (4.77) and (4.79)). We get

$$\frac{4\pi \kappa}{c^2} \left(\varepsilon + \frac{3p}{c^2} \right) = -\frac{4\pi \kappa}{c^2} \left(3\frac{p_\Lambda}{c^2} + \varepsilon_\Lambda \right) - \frac{3\mathbf{a}_r''}{c^2 \mathbf{a}_r}$$

$$= (3\beta - 1)\frac{4\pi \kappa}{c^2} \varepsilon_\Lambda - \frac{3\mathbf{a}_r''}{c^2 \mathbf{a}_r}. \tag{4.102}$$

For the second equation in (4.61) we obtain

$$\frac{4\pi \kappa}{c^2} \left(\varepsilon + \frac{p}{c^2} \right) = -\frac{4\pi \kappa}{c^2} \left(\frac{p_\Lambda}{c^2} - \varepsilon_\Lambda \right) + \frac{1}{c^2} \left(\frac{\mathbf{a}_r'}{\mathbf{a}_r} \right)^2 - \frac{1}{c^2} \left(\frac{\mathbf{a}_r''}{\mathbf{a}_r} \right)$$

$$= (\beta - 1)\frac{4\pi \kappa}{c^2} 2\varepsilon_\Lambda + \frac{1}{c^2} \left(\frac{\mathbf{a}_r'}{\mathbf{a}_r} \right)^2 - \frac{1}{c^2} \left(\frac{\mathbf{a}_r''}{\mathbf{a}_r} \right). \tag{4.103}$$

Resolving (4.102) for $\frac{\mathbf{a}_r''}{\mathbf{a}_r}$ and substituting it into (4.103) leads to

$$\overbrace{\frac{4\pi\kappa}{c^2}\left(\varepsilon + \frac{p}{c^2}\right)} = (\beta - 1)\frac{4\pi\kappa}{c^2}\varepsilon_\Lambda + \frac{1}{c^2}\left(\frac{\mathbf{a}_r'}{\mathbf{a}_r}\right)^2 + \frac{4\pi\kappa}{c^2}\frac{1}{3}\left(\varepsilon + \frac{3p}{c^2}\right) - (3\beta - 1)\frac{4\pi\kappa}{3c^2}2\varepsilon_\Lambda.$$

(4.104)

Resolving for $\left(\frac{\mathbf{a}_r'}{\mathbf{a}_r}\right)^2$, gives

$$\left(\frac{\mathbf{a}_r'}{\mathbf{a}_r}\right)^2 = \frac{8\pi\kappa}{3}\varepsilon + \frac{8\pi\kappa}{3}\varepsilon_\Lambda.$$

(4.105)

Next, we substitute the solutions for ε_Λ and ε, obtained earlier:

$$\varepsilon_\Lambda = \Lambda\mathbf{a}_r^{3(\beta-1)},$$
$$\varepsilon = \varepsilon_0\mathbf{a}_r^{-3(1+\alpha)},$$

(4.106)

leading to

$$\left(\frac{\mathbf{a}_r'}{\mathbf{a}_r}\right)^2 = \frac{8\pi\kappa}{3}\varepsilon_0\mathbf{a}_r^{-3(1+\alpha)} + \frac{8\pi\kappa}{3}\Lambda\mathbf{a}_r^{3(\beta-1)}.$$

(4.107)

Using $\frac{\mathbf{a}_{r,0}}{\mathbf{a}_r} = (1 + z)$ gives finally, where $\mathbf{a}_{r,0}$ is the radius of the universe in the present,

$$\left(\frac{\mathbf{a}_r'}{\mathbf{a}_r}\right)^2 = \frac{8\pi\kappa}{3}\varepsilon_0\mathbf{a}_{r,0}^{-3(1+\alpha)}(1 + z)^{3(1+\alpha)} + \frac{8\pi\kappa}{3}\Lambda\mathbf{a}_{r,0}^{3(\beta-1)}(1 + z)^{3(\beta-1)}$$
$$= \frac{8\pi\kappa}{3}\left[\varepsilon_0\mathbf{a}_{r,0}^{-3(1+\alpha)}(1 + z)^{3(1+\alpha)} + \Lambda\mathbf{a}_{r,0}^{3(1-\beta)}(1 + z)^{3(\beta-1)}\right].$$

(4.108)

Next we introduce

$$\Omega = \frac{\varepsilon_k}{\varepsilon_{\text{crit}}} = \frac{8\pi\kappa}{3H_0^2}\varepsilon_i$$

with

$$\varepsilon_{\text{crit}} = \frac{3H_0^2}{8\pi\kappa},$$

(4.109)

where $\varepsilon_{\text{crit}}$ is the critical density and ε_k refers to the different density components at the current time, e.g. $k = 1$ refers to Λ and $k = 2$ to the matter component, with a given α. H_0 is the current Hubble constant.

With this and defining $\varepsilon_{0i} = \varepsilon_0 \mathbf{a}_{r,0}^{-3(1+\alpha)}$, $\varepsilon_{\Lambda,0} = \Lambda \mathbf{a}_{r,0}^{3(1-\beta)}$

$$H^2 = \left(\frac{\mathbf{a}_r'}{\mathbf{a}_r}\right) = \frac{8\pi\kappa}{3}\left[\varepsilon_{0i}\,(1+z)^{3(1+\alpha)} + \varepsilon_{\Lambda,0}\,(1+z)^{3(1-\beta)}\right]. \quad (4.110)$$

Taking, as an example, the present epoch, the ratio of the velocity and the present radius is just the Hubble constant. Then, from (4.109) we get

$$H_0^2 = \frac{8\pi\kappa}{3}\left[\varepsilon_0 + \varepsilon_{\Lambda,0}\right] = \frac{8\pi\kappa}{3}\varepsilon_{\text{tot}}, \quad (4.111)$$

where ε_{tot} is the total density.

Using (4.110) and extracting the total density, leads to

$$H^2 = H_0^2\left[\Omega_m\,(1+z)^{3(1+\alpha)} + \Omega_\Lambda\,(1+z)^{3(1-\beta)}\right], \quad (4.112)$$

which is the desired result (4.101). The model just described was first introduced by Friedmann and Lemaitre [6, 7].

A further useful relation is obtained, starting from (4.102) and resolving for $\left(\frac{\mathbf{a}_r''}{\mathbf{a}_r}\right)$. Using the above solutions for the densities and using $p = \alpha\varepsilon$, we obtain

$$\left(\frac{\mathbf{a}_r''}{\mathbf{a}_r}\right) = \frac{4\pi\kappa}{3}\,(3\beta - 1)\,\Lambda\mathbf{a}_r^{3(\beta-1)} - \frac{4\pi\kappa}{3}\,(1 + 3\alpha)\,\varepsilon_0\mathbf{a}_r^{-3(1+\alpha)}. \quad (4.113)$$

Inserting again the dependence on the redshift and the current radius of the universes and introduce the Ω_m and Ω_Λ, gives

$$\left(\frac{\mathbf{a}_r''}{\mathbf{a}_r}\right) = H_0^2\left[-\,(1 + 3\alpha)\,\Omega_m\,(1+z)^{3(1+\alpha)} + (3\beta - 1)\,\Omega_\Lambda\,(1+z)^{3(1-\beta)}\right]. \quad (4.114)$$

We use our result for the function $f(z)$, as given in Exercise 4.12, namely

$$f(z) = (1+z)^{3(1-\beta)} = (1+z)^{3(1+w)}, \quad (4.115)$$

where we made a connection to the notation used in [5]. This again yields the relation $\beta = -w$. Our result states that when the Hubble constant changes in time, there *must* be a deviation from $\beta = 1$, which corresponds due to $\beta = -w$ to $w = -1$, the case of a cosmological constant, independent of time. The deviation from $\beta = 1$ cannot be large when H', the time derivative of the Hubble constant, is small.

4.3 An Oscillating Universe

In this section we will show that within the pc-GR exists also an oscillating universe
as a possible solution. This leads us to some proposals, which are however not more
speculative than other accepted scenarios for the evolution of the universe. The main
point is that interesting predictions can be made about the future of the universe! As
we will see, a new *Weltbild* emerges.

The oscillating universe also has a great philosophical advantage: The problem
about the apparent present thermal equilibrium of the universe does not exist, because
after the big crunch the universe stays thermalized due to its thermalization in the
highly compressed phase, which lead to the expansion phase! The question of increasing entropy will be addressed at the end, showing that in relativistic thermodynamics
[10, 11], the model presented indeed satisfies the second law of thermodynamics.

We start from (4.75), use (4.56) and (4.69), set α equal to zero and arrive at

$$\frac{\mathbf{a}_r''}{\mathbf{a}_r} = -\frac{4\pi\kappa}{3}\left(3\frac{p_\Lambda}{c^2} + \varepsilon_\Lambda\right) - \frac{4\pi\kappa}{3}\frac{\varepsilon_0}{\mathbf{a}_r^3}$$

$$\left(\frac{\mathbf{a}_r'}{\mathbf{a}_r}\right)^2 = -8\pi\kappa\frac{p_\Lambda}{c^2} - 2\frac{\mathbf{a}_r''}{\mathbf{a}_r},$$

$$-\frac{d\varepsilon_\Lambda}{d\mathbf{a}_r} = \frac{3}{\mathbf{a}_r}\left(\frac{p_\Lambda}{c^2} + \varepsilon_\Lambda\right), \qquad (4.116)$$

where we repeat that $\mathbf{a}_r$ is the scale factor, which is related to the radius of the
universe. Today, its value is 1!

Exercise 4.13 (Equation (4.116))

Problem. Verify the equations in (4.116).

Solution. Let us repeat first in this order (4.75), (4.56) and (4.69) and in addition
$k = 0$ and $\alpha = 0$,

$$\mathbf{a}_r'' = -\frac{4\pi\kappa}{3}(3\frac{p_\Lambda}{c^2} + \varepsilon_\Lambda)\mathbf{a}_r - \frac{4\pi\kappa}{3}\varepsilon_0\mathbf{a}_r^{-2},$$

$$\frac{8\pi\kappa}{c^2}\varepsilon = -\frac{8\pi\kappa}{c^2}\varepsilon_\Lambda + \left[\frac{3}{c^2}\left(\frac{\mathbf{a}_r'(t)}{\mathbf{a}_r(t)}\right)^2\right],$$

$$\frac{8\pi\kappa}{c^2}\frac{p}{c^2} = -\frac{8\pi\kappa}{c^2}\frac{p_\Lambda}{c^2} - \left[\left(\frac{\mathbf{a}_r'(t)}{c^2\mathbf{a}_r(t)}\right)^2 + \frac{2\mathbf{a}_r''(t)}{c^2\mathbf{a}_r(t)}\right],$$

$$\frac{d\varepsilon_\Lambda}{dt} = -\frac{d(\ln\mathbf{a}_r^3)}{dt}\left(\frac{p_\Lambda}{c^2} + \varepsilon_\Lambda\right). \qquad (4.117)$$

The last equation in (4.116) is immediately reproduced, using the rules to derive the logarithm, i.e. $\frac{d\ln a_r^3}{da_r} = \frac{3}{a_r}$.

The first equation in (4.116) is obtained, dividing the first equation in (4.117) by a_r

$$\frac{\mathbf{a}_r''}{\mathbf{a}_r} = -\frac{4\pi\kappa}{3}\left(3\frac{p_\Lambda}{c^2} + \varepsilon_\Lambda\right) - \frac{4\pi\kappa}{3}\frac{\varepsilon_0}{\mathbf{a}_r^3}. \qquad (4.118)$$

In order to obtain the second equation in (4.116), we start from the third equation in (4.117), which expresses the pressure in terms of $\mathbf{a}_r$, its velocity and acceleration. In this equation we can set $p = 0$, because the equation of state considered is $\frac{p}{c^2} = \alpha\varepsilon$ and for $\alpha = 0$ (dust) the pressure is zero. This leads to, multiplying by c^2,

$$0 = -8\pi\kappa\frac{p_\Lambda}{c^2} - \left(\frac{\mathbf{a}_r'}{\mathbf{a}_r}\right)^2 - 2\left(\frac{\mathbf{a}_r''}{\mathbf{a}_r}\right). \qquad (4.119)$$

Resolving for $\left(\frac{\mathbf{a}_r'}{\mathbf{a}_r}\right)^2$, leads to the second equation in (4.116).

Using (4.35) the following form for $\frac{p_\Lambda}{c^2}$ and ε_Λ can be used:

$$-\frac{p_\Lambda}{c^2} = \Lambda - \tilde{\alpha}\mathbf{a}_r,$$

$$\varepsilon_\Lambda = \Lambda - \tilde{\beta}\mathbf{a}_r. \qquad (4.120)$$

One can also use an $\mathbf{a}_r^n$ dependence, with n being an integer, but we prefer to keep the discussion as simple as possible and use $n = 1$. The $\tilde{\alpha}$ and $\tilde{\beta}$ are extremely small, such that at the present epoch one can consider $\varepsilon_\Lambda = \Lambda$ and $\frac{p_\Lambda}{c^2} = -\varepsilon_\Lambda$. This is the case for constant dark energy.

The ansatz (4.120) for pressure and density is now substituted into the differential equation (4.116), giving the condition

$$\tilde{\beta} = 3\left(\tilde{\alpha} - \tilde{\beta}\right)$$
$$\rightarrow$$
$$\tilde{\alpha} = \frac{4}{3}\tilde{\beta}. \qquad (4.121)$$

With this, we obtain

$$-\frac{p_\Lambda}{c^2} = \Lambda - \frac{4}{3}\tilde{\beta}\mathbf{a}_r,$$

$$\varepsilon_\Lambda = \Lambda - \tilde{\beta}\mathbf{a}_r. \qquad (4.122)$$

For the acceleration (see (4.116)) this is

$$\frac{\mathbf{a}_r''}{\mathbf{a}_r} = \frac{4\pi\kappa}{3}\left(2\Lambda - 3\tilde{\beta}\mathbf{a}_r\right) - \frac{4\pi\kappa}{3}\frac{\varepsilon_0}{\mathbf{a}_r^3}. \tag{4.123}$$

For small $\mathbf{a}_r$ (beginning of our universe) one can neglect the $\tilde{\beta}$-term and we have the usual model, where the acceleration changes from negative to positive at the point $\mathbf{a}_0 = \left(\frac{\varepsilon_0}{2\Lambda}\right)^{\frac{1}{3}}$. For large $\mathbf{a}_r$, however, the (4.123) approaches $-\frac{4\pi\kappa}{c^2}\tilde{\beta}\mathbf{a}_r$. This happens for **very large** $\mathbf{a}_r$, considering the $\tilde{\beta}$ is **very small**. This situation we call *super-decelerating*, because the deceleration increases by $\mathbf{a}^2$.

In order to assure that the acceleration passes though an intermediate phase of positive acceleration (accelerating universe), we consider the function

$$f(\mathbf{a}_r) = 2\Lambda - 3\tilde{\beta}\mathbf{a}_r - \frac{\varepsilon_0}{\mathbf{a}_r^3}, \tag{4.124}$$

which appears in (4.123). Its derivative is

$$f' = -3\tilde{\beta} + \frac{3\varepsilon_0}{\mathbf{a}_r^4} = 0$$
$$\rightarrow$$
$$\mathbf{a}_{\text{max}} = \left(\frac{\varepsilon_0}{\tilde{\beta}}\right)^{\frac{1}{4}}. \tag{4.125}$$

This we substitute into (4.124), which gives

$$f(\mathbf{a}_{\text{max}}) = 2\Lambda - 4\varepsilon_0^{\frac{1}{4}}\tilde{\beta}^{\frac{3}{4}}. \tag{4.126}$$

Requiring that this is positive, implies

$$\tilde{\beta} < \frac{\Lambda^{\frac{4}{3}}}{2^{\frac{4}{3}}\varepsilon_0^{\frac{1}{3}}}. \tag{4.127}$$

Considering that $\Lambda \ll \varepsilon_0$, this is a **very small** value.

Now we look for the maximum of (4.124) for **very large** $\mathbf{a}_r$. We will do it approximately, because the general equation is of fourth order. For very large $\mathbf{a}_r$ we can neglect the term proportional to ε_0 and this tells us that the acceleration is again zero for approximately

$$\mathbf{a}_r = \mathbf{a}_2 \approx \frac{2}{3}\left(\frac{\Lambda}{\tilde{\beta}}\right). \tag{4.128}$$

Let us look at the velocity (4.105), applying the Eqs. (4.118), (4.120), we obtain

$$\left(\frac{\mathbf{a}_r'}{\mathbf{a}_r}\right)^2 = \frac{8\pi\kappa}{3}\left\{\Lambda - \tilde{\beta}\mathbf{a}_r + \frac{\varepsilon_0}{\mathbf{a}_r^3}\right\}. \tag{4.129}$$

Equation (4.129) is zero for

$$\Lambda - \tilde{\beta}\mathbf{a}_r + \frac{\varepsilon_0}{\mathbf{a}_r^3} = 0, \tag{4.130}$$

which is a fourth order equation. We solve it for very large **a**, neglecting the ε_0-term:

$$\mathbf{a}_r = \mathbf{a}_3 \approx \left(\frac{\Lambda}{\tilde{\beta}}\right), \tag{4.131}$$

which is larger than $\mathbf{a}_2$, were the acceleration turns negative.

In conclusion, the velocity of the expansion is zero at $\mathbf{a}_3$, while the acceleration is negative (deceleration). Thus, $\mathbf{a}_3$ represents a turning point and the universe starts to contract again.

It is interesting to see what the equation of state for the dark energy is, within this model. For that, we calculate the ratio

$$\frac{\frac{p_\Lambda}{c^2}}{\varepsilon_\Lambda} = \frac{\Lambda - \frac{4}{3}\tilde{\beta}\mathbf{a}_r}{\Lambda - \tilde{\beta}\mathbf{a}_r}$$

$$\rightarrow$$

$$\frac{p_\Lambda}{c^2} = g(\mathbf{a}_r)\varepsilon_\Lambda = w\varepsilon_\Lambda$$

$$\text{with } g(\mathbf{a}_r) = -\left(\frac{\Lambda - \frac{4}{3}\tilde{\beta}\mathbf{a}_r}{\Lambda - \tilde{\beta}\mathbf{a}_r}\right), \tag{4.132}$$

which gives a relation of the proportionality factor w in the equation of state to $\mathbf{a}_r$. The function $g(\mathbf{a}_r)$ produces an equation of state where for small $\mathbf{a}_r$ the pressure is proportional to minus the density. However, at $\mathbf{a}_r = \frac{3\Lambda}{4\tilde{\beta}}$ this function is zero, i.e. the pressure is zero, and at $\mathbf{a}_3$ (see (4.131)), the point where the expansion of the universe is reversed, the function approaches infinity, which might be due to the approximation made. This is a strange equation of state but there is, at the moment, no experimental evidence that it is wrong!

The main point is that exotic assumptions can be made, where the universe is oscillating, as exotic as other solutions one can encounter in the literature. The above consideration refer *only* to one part of the oscillation, namely the time from the beginning toward the end. What happens at the birth and at the crush of the universe can not be deduced. At these points, first the effects of the minimal length have to be taken into account explicitly and second quantum effects have to be included, which is out of the scope in the present form of the theory. Thus, no definite prediction of the fate of the universe can be made! The question is which scenario is more attractive.

This depends on the philosophical view of each scientist. Of course, the question is also: Which scenario is real? Up to now, no definite answer can be made.

Next, we determine the history of the pressure and density of the dust, which serves as a consistency check. In this case, the pressure is given in (4.55), which depends on the velocity (4.129) and on their acceleration (4.123) (k is set to zero).

Using for $\frac{p_\Lambda}{c^2}$ and ε_Λ, as given in (4.122), substituting them into (4.56), we obtain step by step the following expression:

$$
\begin{aligned}
\frac{8\pi\kappa}{c^2}\frac{p}{c^2} &= \frac{8\pi\kappa}{c^2}\frac{p}{c^2} - \frac{1}{c^2}\left(\frac{a_r'}{a_r}\right)^2 - \frac{2}{c^2}\frac{a_r''}{a_r} \\
&= -\frac{8\pi\kappa}{c^2}\left(\Lambda - \frac{4}{3}\tilde{\beta}a_r\right) - \frac{1}{2c^2}\frac{8\pi\kappa}{3}\left(\Lambda - \tilde{\beta}a_r + \frac{\varepsilon_0}{a_r^3}\right) \\
&\quad - \frac{2}{c^2}\left(-\frac{4\pi\kappa}{c^2}\right)\left[3\left(\Lambda - \frac{4}{3}\tilde{\beta}a_r\right) + \left(\Lambda - \tilde{\beta}a_r\right) + \frac{\varepsilon_0}{a_r^3}\right] \\
&= 0.
\end{aligned}
\tag{4.133}
$$

This results was to be expected, because in the equation of state for the matter we did set $\alpha = 0$.

Using the same previous steps and equations, we obtain for the density

$$
\begin{aligned}
\frac{8\pi\kappa}{c^2}\varepsilon &= \frac{8\pi\kappa}{c^2}\varepsilon_\Lambda + \frac{3}{c^2}\left(\frac{a_r'}{a_r}\right)^2 \\
&= \frac{8\pi\kappa}{c^2}\left(\Lambda - \tilde{\beta}a_r\right) + \frac{3}{c^2}\frac{8\pi\kappa}{3}\left(\Lambda - \tilde{\beta}a_r + \frac{\varepsilon_0}{a_r^3}\right) \\
&\quad - \frac{8\pi\kappa}{c^2}\frac{\varepsilon_0}{a_r^3}.
\end{aligned}
\tag{4.134}
$$

This result is to be expected because it is consistent with (4.74).

In the present epoch $a_r = 1$, thus $\varepsilon = \varepsilon_0$, as it was defined. (4.134) tells us that in the early epoch the density was very large, with a singularity at $a_r = 0$. This is, of course, not physical, because quantum effects will set in; these are not considered here. For increasing a_r the density decreases and approaches very small values. In the above considerations on the dark energy and pressure we have neglected the mass term for very large a_r. In fact, there have to be corrections due to the $\frac{\varepsilon_0}{a_r^3}$-term and ε_Λ does not vanish at a_3. Thus, the singularity is the result of the approximation. This finding is also consistent, if we assume that there is a coupling between the mass and the dark energy distribution: *When the mass density is very small, then also the dark energy gets very small.*

One may ask, what the explicit time dependence of the radius of the universe is. For that, we start from (4.129), take the square root and multiply by $\mathbf{a}_r$:

$$\mathbf{a}'_r = \sqrt{\frac{8\pi\kappa}{3}}\,\mathbf{a}_r\sqrt{\Lambda - \tilde{\beta}\mathbf{a}_r + \frac{\varepsilon_0}{\mathbf{a}_r^3}}\,. \tag{4.135}$$

Applying a separation of variables gives the equation

$$\frac{d\mathbf{a}_r}{\mathbf{a}_r\sqrt{\Lambda - \tilde{\beta}\mathbf{a}_r + \frac{\varepsilon_0}{\mathbf{a}_r^3}}} = \sqrt{\frac{8\pi\kappa}{3}}\,dt, \tag{4.136}$$

which is easily rewritten as

$$T - T_i = \sqrt{\frac{3}{8\pi\kappa}}\int_{\mathbf{a}_{ri}}^{\mathbf{a}_r}\frac{d\mathbf{a}_r}{\mathbf{a}_r\sqrt{\Lambda - \tilde{\beta}\mathbf{a}_r + \frac{\varepsilon_0}{\mathbf{a}_r^3}}}. \tag{4.137}$$

The T denotes the time of the present epoch, T_i when the integration starts in the past (for example $T_i = 0$), $\mathbf{a}_r$ is the present radius of the universe while $\mathbf{a}_{ri}$ its initial value (for example, when $T_i = 0$ then $\mathbf{a}_{ri} = 0$).

This integral can be put into a more convenient form, namely

$$T - T_i = \frac{\sqrt{6}}{c}\int_{\mathbf{a}_{ri}}^{\mathbf{a}_r}\frac{\sqrt{\mathbf{a}_r}\,d\mathbf{a}_r}{\sqrt{\Lambda\mathbf{a}_r^3 - \tilde{\beta}\mathbf{a}_r^4 + 2C}}. \tag{4.138}$$

This is an integral which can only be solved numerically.

4.3.1 The Adiabatic Expansion and Contraction of the Universe

This subsection serves to give a short resumé on the thermodynamical properties of the oscillating universe in our theory. An excellent introduction to Thermodynamics in Special Relativity and General Relativity can be found in [10, 11] and we refer the reader to it for deeper understanding.

In the work of Friedmann and Tolman [6, 10–12] the possibility of oscillatory universes were discussed. As shown in [10–12] strictly periodic universes, were the velocity of expansion at minimal and maximal radius is zero, do not exist and leads to a contradiction. However, *quasi-periodic* solutions do exists, where the velocity towards minimal radius becomes in the theory of Friedmann and Tolman singular (in our theory this singularity will be avoided due to the appearance of a minimal length). The last statement is in accordance to the previous discussion, i.e. in our model the velocity becomes singular at small values of $\mathbf{a}_r$. In [10–12] a relativistic thermodynamics was developed, showing that in certain models the expansion and contraction is *adiabatic*, thus the total number of entropy does not increase, though it might appear so for a local observer. The reason is that there is a contribution of the metric, due to the expansion (contraction) of the universe, which compensates apparent increases of the entropy.

The detailed reason is as follows: The local change of entropy satisfies the equation [11]

$$\frac{d}{dt}(s\delta V) \geq 0, \tag{4.139}$$

where s is the entropy density and δV the small volume accessible for the measurements of an observer. In case of an *adiabatic* change this is equal to zero. Setting δV constant is equivalent to saying that the entropy in a given volume can only increase or being constant, the latter for an adiabatic change. This is what is usually observed. However, as stated by R.C. Tolman, the volume itself depends on the metric! The above equation can formally be rewritten as

$$\delta V \frac{d}{dt} s \geq -s\frac{d}{dt}\delta V. \tag{4.140}$$

For $\delta V = 0$ this is just the classical statement that the entropy can only increase or stay constant.

Restricting now to our specific model with the length element, as given in (4.1) with (4.46), the volume element is given by

$$\delta V = r^2 \mathbf{a}_r^3 \sin\vartheta\, \delta r \delta\vartheta \delta\varphi, \tag{4.141}$$

with $\mathbf{a}_r^3 = e^{\frac{3}{2}g}$, as defined in (4.46). The metric contribution is the $\mathbf{a}_r^3$ factor. Due to this dependence, *the entropy density may increase, compensated by a change in the metric*. In such a universe, the expansion and contraction is adiabatic with no net change of the entropy!

Let us see if our model does satisfy these assumptions. For that we start from (4.139), using the equal sign for an adiabatic change, with the volume element given by (4.140). The only time- ($\mathbf{a}_r$-)dependent factor is $\mathbf{a}_r^3$. Thus, we can extract the other factors leading to

$$\frac{d}{dt}\left(s\mathbf{a}_r^3\right) = 0. \tag{4.142}$$

The final step is to realize that for an non-interacting gas with an equation of state of the form $\frac{p}{c^2} = \alpha\varepsilon$, the entropy will be proportional to the number of particles, thus the entropy density is proportional to the mass density ($s \sim \varepsilon$). The proportionality factor depends on numbers and natural constants. Extracting these constant factors we finally arrive at

$$\frac{d}{dt}\left(\varepsilon \mathbf{a}_r^3\right) = 0. \tag{4.143}$$

This is nothing but (4.66) with the right hand side set to zero and the pressure $p = 0$ (dust). This is the model we have discussed above in which, obviously, Eq.(4.143) is satisfied! Thus, our model presents a universe which expands and contract *adiabatically*!

Of course, the model does not take into account the presence of radiation, which should, according to [10, 12] still lead to an oscillatory universe. The main approximation is the isotropy of the universe, i.e., local variations are excluded, which finally might -or might not- lead to a net increase of entropy. The main point here is to state that a universe still may expand and contract adiabatically, i.e. being a periodic universe, without running into problems with the increase of entropy.

However, because irreversible processes take place locally, the total entropy of the universe has to increase. This is also discussed in [10, 11] and deduced that each posterior cycle takes longer. Inversely, each previous cycle takes a shorter time, such that at one point one reduces the problem again to a beginning of the universe. A possible alternative is to investigate how so-called irreversible processes, as the production of elements, can be turned into be reversible, which means that in a final crush of the universe all elements dissociate again into protons and electrons. We believe that such a formulation should exist.

A more involved model was published in [13, 14], including matter, radiation and the contribution of the (phenomenological) dark energy. The latter is described via a scalar field and an interaction between matter and the scalar field is proposed. Scalar fields are identified with the dark energy, i.e., the scalar field Lagrangian describes the dynamics of the dark energy (in our theory we treat it via a classical fluid and not through a quantum field). This is particular interesting in our case, because we belief that the microscopic origin probably lies in a coupling of matter with vacuum fluctuations. Though, the model in [13] does not have necessarily a connection to string theory, the authors include a discussion of deeper origin where string theory enters. The final model is the one of two world branes which periodically collide. In [13] the entropy is also maintained constant, i.e. the expansion and contraction is adiabatic. Also here, the inclusion of real irreversible processes, like changes in the chemical composition as argued in [10, 11], are not considered. It is interesting to us that in [14] the equation of state variable w also is near to -1 in the present epoch, while for later times it is increasing, as in our theory.

There is a similarity of our theory to these early models, because both include the dark energy. We did not treat the dark energy as a quantum field but as a classical fluid. However, the coupling of matter to the dark energy is implicitly included (see Chap. 2 on the general formulation of pc-GR) by letting the dark energy density approach zero for very dilute mass density and increase for larger mass density.

A periodically universe in pc-GR is tremendously interesting: There is never a beginning of the world and never an end. There is no big bang, only big-bang-like epochs within the vibrating universe. The question of *what* is/was before the big bang is irrelevant. The universe was always there; it always existed! No singularity is present. Indeed, we have thus created a new *Weltbild*, quite beautiful!

References

1. P.O. Hess, L. Maghlaoui, W. Greiner, The Robertson-Walker metric in a pseudo-complex general relativity. Int. J. Mod. Phys. E **19**, 1315 (2010)
2. R. Adler, M. Bazin, M. Schiffer, *Introduction to General Relativity*, 2nd edn. (McGraw Hill, New York, 1975)
3. S. Carroll, *Spacetime and Geometry. An Introduction to General Relativity* (Addison-Wesley, San Francisco, 2004)
4. P.O. Hess, W. Greiner, Pseudo-complex general relativity. Int. J. Mod. Phys. E **18**, 51 (2009)
5. P.J.E. Peebles, The cosmologiocal constant and dark energy. Rev. Mod. Phys. **75**, 559 (2003)
6. A. Friedmann, Über die Krümmung des Raumes. Z. Phys. **10**, 377 (1922)
7. G. Lemaître, L'Univers en expansion. Ann. Soc. Sci. Bruxelles A **53**, 51 (1933). Reprinted as a Golden Oldie. Gen. Relativ. Gravit. **29**, 637 (1997)
8. W. Greiner, L. Neise, H. Stöcker, *Thermodynamics and Statistical Mechanics* (Springer, Heidelberg, 1995)
9. A. González, F.S. Guzmáni, Accretion of phantom scalar field into a black hole. Phys. Rev. D **79**, 121501(R) (2009)
10. R.C. Tolman, On the problem of entropy of the universe as a whole. Phys. Rev. **37**, 1639 (1931)
11. R.C. Tolman, *Relativity, Thermodynamics and Cosmology* (Oxford University Press, New York, 1934)
12. R.C. Tolman, On the theoretical requirement for a periodic behaviour of the universe. Phys. Rev. **38**, 1758 (1931)
13. P.J. Steinhardt, N. Turok, A cyclic model of the universe. Science **296**, 1436 (2002)
14. P.J. Steinhardt, N. Turok, Cosmic evolution in a cyclic universe. Phys. Rev. D **65**, 126003 (2003)

Chapter 5
Observational Verifications of pc-GR

In the previous chapters, the pc-solution for a non-rotating dark star (Schwarzschild), a charged dark star (Reissner-Nordström) and a rotating dark star (Kerr) were discussed. The theory was applied to the Robertson-Walker model of the universe, encountering new solutions, like a finite or vanishing acceleration of the Universe for very large times and an oscillating universe.

It is important to investigate what are the predictions of pc-GR and how they can be experimentally verified. This is what we address now:

Especially in the last decades a lot of progress has been achieved in radioastronomy, working in electromagnetic wave lengths which penetrates easily the dust in our and other galaxies. At the very beginning of radioastronomy, one of the main difficulties was the poor angular resolution. This was overcome by the *Baseline Interferometry*, where two or more radio dishes are connected by, e.g., coaxial optical cables, joining the measured signals. In such an array of antennas one has to take into account the spatial distance to each other and measuring the time differences between two points as exactly as possible. With this, several radio antennas behave as one huge antenna, the size of the array.

In the present, one has developed the *Very Long Baseline Interferometry* (VLBI), where radio antennas around the whole world are connected. The time is measured, using local atomic clocks. Antennas in North- and South-America and Asia are connected through huge distances, forming an effective radio antenna of the size of the Earth! ALMA (*Atacama Large Millimeter/Submillimeter Array* [1]) in the Atacama desert in Chile is one of largest local arrays and has recently started observations. The angular resolution to be achieved is about $8\,\mu$as, enough to resolve the supposed black hole in the center of our galaxy, with about 4 million solar masses, and the super-massive one in the active galaxy M87. This gives hope to directly observe the accretion disk around M87.

© Springer International Publishing Switzerland 2016
P.O. Hess et al., *Pseudo-Complex General Relativity*,
FIAS Interdisciplinary Science Series, DOI 10.1007/978-3-319-25061-8_5

The massive object in the center of our galaxy (in Sgr A*) seems not to have an accretion disk, which is the reason that it is not active. However, a dense gas cloud has been detected falling into the aggregation zone of the Galactic Center [2]. The gas will be attracted by the Galactic Center from the second half of 2013 on and as the lower limit of the absolute value of the black hole spin is larger than $0.5m$. This will allow for additional tests of theories in the strong field limit.

There is another puzzle: One observes at centers of active galaxies *Quasi Periodic Oscillations* (QPO) [3–5]. These are thought to be local emissions in the accretion disk, moving in an orbit around the central mass. Such QPO's are also observed in galactic "black holes" [6–10]. There, the situation is more complicated, due to the presence of a stellar partner, providing mass to the accretion disk. The advantage of these so-called *galactic black holes* is that the frequency and Fe-K lines are observed simultaneously! *If one assumes* that the QPO's are local emissions in the accretion disk, orbiting around the central mass (as in the center of active galaxies), then this permits to deduce the distance to the center using the predicted dependence of the orbital frequency on r. The Fe-K lines permit to deduce the redshift, which is given in a theory also as a function on the radial distance. In a consistent theory, the r deduced from the QPO's frequency and from the redshift have to agree. And there is the problem: *They do not agree*! One possible solution is to assume an influence of the stellar partner on the accretion disk, provoking oscillations within the disk [11]. However, not all astronomers accept this interpretation, but rather assume that the physics of the accretion disk around large masses at the center of a galaxy is the same for galactic "black holes".

As one can see, astronomy has developed in the last decades from an purely observational science with large errors to an exact science with extremely accurate measurements. This is the reason why we have to calculate observational consequences of our theory and compare it to the standard GR. In the near future gravitational theories will be able to be tested in extremely strong gravitational fields. Tests of GR are numerous for weak gravitational fields [12], i.e. all tests in the solar system (e.g., perihelion shifts, frame dragging and time measurements (GPS)) or, one of the best known, the Hulse-Taylor pulsar [13], which gave the first indirect hint to gravitational waves, but is still in the weak gravitational limit.

This chapter is, therefore, dedicated to make definite predictions of pc-GR and compare them to standard GR, in very strong gravitational fields. First, the orbital motion of a particle in a circular orbit is discussed, simulating the motion of a QPO. Also the redshift is calculated and we show that pc-GR can reconcile the observation with theory, under the assumption that QPO's are local emissions moving with the accretion disk and are not oscillations provoked by a stellar partner. Because the first observation to be published will be probably a picture of the accretion disk, we also reconstruct the images of such a disk using the *raytracing method* [14, 15]. The name has its origin in the method, where a light ray is *traced back* from the camera of the observer to the point of origin. The method will be discussed in more detail in the corresponding section.

5.1 Motion of a Particle in a Circular Orbit: Orbital Frequency and Redshift

The Kerr-metric is the most relevant one, because the probability for high spin for large masses is great. This is due to the fact that even when the original mass before collapse has a low rotational frequency, when concentrated in a small volume this frequency increases.

First, the pc-Kerr metric is resumed. The detailed derivation can be found in Chap. 3. We have respectively

$$
\begin{aligned}
g_{00} &= -\frac{r^2 - 2mr + a^2 \cos^2 \vartheta + \frac{B}{2r}}{r^2 + a^2 \cos^2 \vartheta}, \\
g_{11} &= \frac{r^2 + a^2 \cos^2 \vartheta}{r^2 - 2mr + a^2 + \frac{B}{2r}}, \\
g_{22} &= r^2 + a^2 \cos^2 \vartheta, \\
g_{33} &= (r^2 + a^2) \sin^2 \vartheta + \frac{a^2 \sin^4 \vartheta \left(2mr - \frac{B}{2r}\right)}{r^2 + a^2 \cos^2 \vartheta}, \\
g_{03} &= \frac{-a \sin^2 \vartheta \, 2mr + a\frac{B}{2r} \sin^2 \vartheta}{r^2 + a^2 \cos^2 \vartheta},
\end{aligned}
\tag{5.1}
$$

where a is he spin-parameter, describing the spin of the central object (see Chap. 3). For $a = 0$ the Kerr metric reduces to the pseudo-complex Schwarzschild solution.

5.1.1 Radial Dependence of the Angular Frequency and Stable Orbits

Instead of describing explicitly the whole accretion disk, we will for the moment simulate the motion of one particle within this disk on a circular orbit. Later, the physics of the accretion disk will be discussed, including the emission of light from there.

The first observable is the orbital frequency, which can be measured probably in near future [2]. Flares are already observed at the galactic center black hole candidate Sgr A* [3, 4].

To determine the angular velocity, one proceeds similar as in [16, 17]. The Lagrange function is given by

$$
L = g_{00}c^2\dot{t}^2 + g_{11}\dot{r}^2 + g_{22}\dot{\vartheta}^2 + g_{33}\dot{\varphi}^2 + 2g_{03}c\dot{t}\dot{\varphi} = \frac{ds^2}{ds^2} = -1,
\tag{5.2}
$$

with $\dot{r} = \frac{dr}{ds}$ and similar for the other variables. The s represents the curve parameter, where the dot represents $\frac{d}{ds}$. The variation of L yields the geodesic equations from which only the radial one is of interest, namely

$$\frac{d}{ds}(2g_{11}\dot{r}) = g_{00}'c^2\dot{t}^2 + g_{11}'\dot{r}^2 + g_{22}'\dot{\vartheta}^2 + g_{33}'\dot{\varphi}^2 + 2g_{03}'ci\dot{\varphi}. \tag{5.3}$$

The prime denotes the derivative $\frac{\partial}{\partial r}$. For simplicity, the calculations are restricted to circular orbits in the equatorial plane, i.e. $r = r_0$, $\dot{r} = 0$, $\vartheta = \frac{\pi}{2}$ and $\dot{\vartheta} = 0$. Equation (5.3) becomes

$$0 = g_{00}'(r_0)c^2\dot{t}^2 + g_{33}'(r_0)\omega^2\dot{t}^2 + 2g_{03}'(r_0)\omega c\dot{t}^2, \tag{5.4}$$

where the angular frequency $\omega = \frac{d\varphi}{dt} = \frac{\dot{\varphi}}{\dot{t}}$ has been introduced.

The quadratic equation (5.4) has two solutions:

$$\omega_{\pm} = -c\frac{g_{03}'}{g_{33}'} \pm c\sqrt{\frac{\left(g_{03}'\right)^2 - g_{00}'g_{33}'}{g_{33}'^2}}. \tag{5.5}$$

Inserting the Kerr metric (5.1) yields

$$\omega_{\pm} = c\frac{-ah(r) \pm \sqrt{2rh(r)}}{2r - a^2h(r)} = \frac{c\sqrt{h(r)}}{a\sqrt{h(r)} \pm \sqrt{2r}}, \tag{5.6}$$

with

$$h(r) = \frac{2m}{r^2} - \frac{3B}{2r^4}. \tag{5.7}$$

Exercise 5.1 (Angular frequency)

Problem. Proof (5.6) with (5.7).

Solution. We start from (5.4) and divide it by $\dot{t}^2$. The resulting equation is

$$g_{33}'\omega^2 + 2g_{03}'\omega c + g_{00}'c^2$$
$$= \omega^2 + 2\frac{g_{03}'}{g_{33}'}\omega c + \frac{g_{00}'}{g_{33}'}c^2 = 0, \tag{5.8}$$

where in the last step we divided also by g_{33}'. This is a quadratic equation whose solution is (5.5).

Now, we insert the Kerr-metric, given in (5.1). We do it in steps, first expressing the metric components, which appear in the formula of $\omega_{\pm}$, in terms of r. The ϑ is set to $\frac{\pi}{2}$ (orbital plane). We have

$$g_{00} = -\left(1 - \frac{2m}{r} + \frac{B}{2r^3}\right),$$

$$g_{03} = -a\left[\frac{2m}{r} - \frac{B}{2r^3}\right],$$

$$g_{33} = (r^2 + a^2) + a^2\left(\frac{2m}{r} - \frac{B}{2r^3}\right). \qquad (5.9)$$

The derivatives with respect to r are

$$g'_{00} = -\frac{2m}{r^2} + \frac{3B}{2r^4} = -h(r),$$

$$g'_{03} = a\left[\frac{2m}{r^2} - \frac{3B}{2r^4}\right] = ah(r),$$

$$g'_{33} = 2r - a^2\left(\frac{2m}{r^2} - \frac{3B}{2r^4}\right) = 2r - a^2h(r), \qquad (5.10)$$

where also the definition of $h(r)$ is given.

In the square root of (5.5), extracting g_{33}^2 is given by $-g_{33}$, because g_{33} is positive. Substituting this into the solution (5.5), we obtain

$$
\begin{aligned}
\omega_\pm &= c\frac{\left[-ah(r) \pm \sqrt{a^2h^2(r) + 2rh(r) - a^2h^2(r)}\right]}{2r - a^2h(r)} \\
&= c\frac{\left[-ah(r) \pm \sqrt{2rh(r)}\right]}{2r - a^2h(r)} \\
&= -c\frac{\left[ah(r) \mp \sqrt{2rh(r)}\right]}{\left[a\sqrt{h(r)} + \sqrt{2r}\right]\left[-a\sqrt{h(r)} + \sqrt{2r}\right]} \\
&= \pm c\sqrt{h(r)}\frac{\sqrt{2r} \mp a\sqrt{h(r)}}{\left[a\sqrt{h(r)} + \sqrt{2r}\right]\left[-a\sqrt{h(r)} + \sqrt{2r}\right]} \\
&= -c\sqrt{h(r)}\frac{\left[a\sqrt{h(r)} \mp \sqrt{2r}\right]}{\left[a\sqrt{h(r)} + \sqrt{2r}\right]\left[-a\sqrt{h(r)} + \sqrt{2r}\right]} \\
&= \frac{c\sqrt{h(r)}}{\left[a\sqrt{h(r)} \pm \sqrt{2r}\right]}. \qquad (5.11)
\end{aligned}
$$

For $h(r) > 0$, (5.6) displays two real solutions, one for co-rotation (which orbits in the same direction as the central body rotates, also called *prograde*) and one for counter–rotation (which orbits in the opposite direction as the central object rotates, also called *retrograde*) with respect to the rotating central body. Note, that the condition $h(r) > 0$ is equivalent to require $r^2 > (3B)/(4m)$, which for $B = B_{\min} = \frac{64}{27} m^3$, implies $r > (4/3) m$. In Exercise 5.2 one has to show which sign of a corresponds to the prograde and retrograde orbit.

Exercise 5.2 (Prograde and retrograde orbital motion)

Problem. Show that ω_+ corresponds to the prograde and ω_- to the retrograde orbital motion.

Solution. The angular momentum of the central mass is given by J, which can have a positive or a negative sign, were the positive sign is usually chosen. For convenience we say that the angular momentum for this rotation points towards the north pole of the central mass. One has to use the connection between the parameter a and the angular momentum J as given in [16, 17], namely

$$a = \frac{\kappa J}{mc^3}. \tag{5.12}$$

The J is chosen positive and accordingly the a is positive. Inspecting (5.6), the sign of ω is determined by the denominator (the numerator is always positive). Since $h(r) > 0$, for ω_- the denominator is $a\sqrt{h(r)} - \sqrt{2r}$, which is always lower in its absolute value than $a\sqrt{h(r)} + \sqrt{2r}$, which is the denominator for ω_+. Thus $|\omega_-| > |\omega_+|$ always holds.

Consider now the limit of large radial distances. There, the frame dragging effect is small and the prograde and retrograde orbits are well defined. In this limit $2r > a^2 h(r)$ always holds. Then, ω_+ is positive and ω_- is negative. Consequently ω_+ describes prograde orbits, whereas ω_- describes retrograde orbits. The clear identification of prograde and retrograde orbits vanish for small radial distances. There one has $2r < a^2 h(r)$. In this case ω_+ and ω_- are both positive, which means that the particle in the formerly retrograde orbit *gets dragged* and can rotate in the same direction as the central object (prograde).

In Fig. 5.1 the angular frequency ω_+ of a mass in a co-rotation (prograde) orbit is plotted versus the radial distance, the rotational parameter $a = 0.995 \, m$. In Fig. 5.2 the same is shown for a retrograde orbit. For such a large value of a the last stable orbit ends at a fixed value, given by the position of the star's surface (this will be shown further below when stable orbits are discussed in general). The radial distance is given in units of m, and ω in units of c/m. For a mass of 4 millions times the mass of the sun [4], as is the case for the object at the center of our galaxy, a value of 0.219, at the maximum of $\omega = 2\pi \nu$ corresponds to a orbital period of 9.4 min. In Fig. 5.1 the frequency shows a maximum below $r = 2$, which is below the Schwarzschild radius.

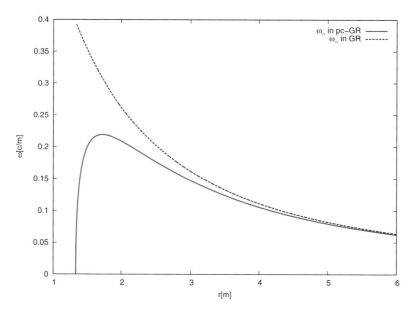

Fig. 5.1 Orbital frequency as a function of r, for stable geodesic prograde *circular motion*. The value $\omega = 0.219$, for a mass of 4 million suns of the *dark star*, corresponds to about 9.4 min for a *full circle*. The plot is done for parameter values of $a = 0.995\,m$ and $B = \frac{64}{27}\,m^3$

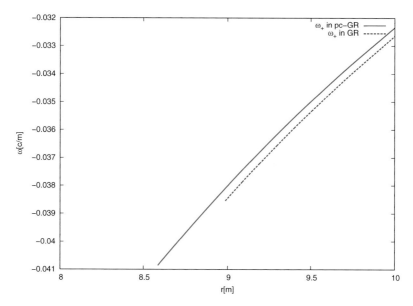

Fig. 5.2 Orbital frequency as a function of r, for retrograde circular motion. The parameters for this plot are $a = 0.995m$ and $B = \frac{64}{27}\,m^3$

The reason for the maximum can be found in (5.5): The frequency depends on the *derivative* of the metric. In the Kerr case this derivative increases continuously toward the event horizon, which implies that ω always increases toward smaller radial distances. In pc-Kerr, however, there is a minimum in the metric coefficient g_{00}, below the Schwarzschild radius, which requires that the derivative of the metric coefficients have a *turning point*, before reaching the absolute minimum, from which on it decreases again. This turning point generates the maximum.

A particular form of (5.6) is given by

$$\omega_{\pm} = \frac{c}{a \pm \sqrt{\frac{2r}{h(r)}}}, \tag{5.13}$$

from which one can deduce that the position of this maximum in the frequency is *independent of the value of a*. This is an important finding, because once a maximum in ω_- is observed, *the position of this maximum together with the mass will determine B*. We also refer to the section on the simulation of the accretion disk, later in this chapter. There, a dark ring is predicted (see section on raytracing), which is intimately related to the position of this maximal frequency. Thus, the position of the maximum should be clearly observable.

Exercise 5.3 (Maximal angular frequency for prograde rotation)

Problem. Determine the position of the maximum of ω_-, for prograde rotation.

Solution. The maximum of ω_+ can be deduced starting from (5.6), calculating its derivative. We resume Eq. (5.6) and give its derivative:

$$\omega_+ = \frac{c\sqrt{h(r)}}{a\sqrt{h(r)} + \sqrt{2r}},$$

$$\omega'_+ = \frac{ch'}{2\sqrt{h}\left[a\sqrt{h} + \sqrt{2r}\right]} - c\sqrt{h}\frac{\left[a\frac{h'}{2\sqrt{h}} + \frac{1}{\sqrt{2r}}\right]}{\left[a\sqrt{h} + \sqrt{2r}\right]^2}$$

$$= \frac{\left[\frac{ch'}{2\sqrt{h}}\sqrt{2r} - \frac{c\sqrt{h}}{\sqrt{2r}}\right]}{\left[a\sqrt{h} + \sqrt{2r}\right]^2}. \tag{5.14}$$

The derivative has to be zero, which is satisfied by

$$h' = \frac{h}{r}. \tag{5.15}$$

The function h is given in (5.7) and its derivative is

$$h' = -\frac{4m}{r^3} + \frac{6B}{r^5},$$

(5.16)

which gives for (5.15)

$$r = \sqrt{\frac{5B}{4m}}.$$

(5.17)

Using for B its minimal value $\frac{64}{27} m^3$, gives

$$r = \sqrt{\frac{80}{27}} \, m \approx 1.72 \, m,$$

(5.18)

in excellent agreement to the numerical value in Fig. 5.1.

Noticeable differences to standard GR appear around the Schwarzschild radius and starting from near twice this value. The main feature for the orbital time of a particle is that it takes more time to circle the center than in GR. In both theories one obtains the angular frequency as a function of the radial distance, which are different near the Schwarzschild radius. In pc-GR there are two solutions, namely one for small r and one for large r, while in GR there is always one solution, which agrees more with the solution at large r in pc-GR. Therefore, in order to distinguish between both theories *a second observable is needed*. This second observable is the *redshift*, whose function in r will be different in both theories. A consistent description is obtained, when in a theory *both values of r*, one deduced in adjusting the orbital frequency and the other adjusting the redshift, *result in the same r*.

As will be shown in the section on stable orbits, a stable orbit for *prograde* motion and $a \geq 0.4$ always exists. In the case of *retrograde* orbits we do not expect to see big differences as a last stable orbit exists at $r > 8$ (for $a = 0.995 \, m$). Below that value there are no stable orbits anymore. For such large radial distances, there is no detectable difference between GR and pc-GR. Thus, we restrict to prograde orbits.

5.1.1.1 General Orbits

Up to now, only stable orbits were considered, However, one can also derive constraints on the orbital frequencies for more general orbits. Following [18] we require that the line element

$$ds^2 = g_{00}c^2dt^2 + g_{11}dr^2 + g_{22}d\vartheta^2 + g_{33}d\varphi^2 + 2g_{03}cdtd\varphi$$

(5.19)

is negative.

The difference to the former discussion is that the Lagrangian of (5.2) was varied in order to obtain the geodesic equation, while now this is not the case any more. Again, we restrict the calculations to circular motions in the equatorial plane ($dr = 0, d\vartheta = 0, \vartheta = \frac{\pi}{2}$). The limiting case $ds^2 = 0$ corresponds to a *circular rotating photon*. This will produce two limiting lines between where the solutions for massive particles lie. For a photon, we get

$$g_{00}c^2 dt^2 + g_{33}\omega^2 dt^2 + 2g_{03}c\omega dt^2 = 0, \qquad (5.20)$$

where again $\omega = \frac{d\varphi}{dt}$ is used. This is a quadratic equation with the solution

$$\bar{\omega}_{\pm} = -c\frac{g_{03}}{g_{33}} \pm c\sqrt{\frac{(g_{03})^2 - g_{00}g_{33}}{g_{33}^2}}. \qquad (5.21)$$

This seems to be equal to (5.5), but one clearly notes that instead of the derivatives of the metric now the metric without the derivatives appear.

Inserting (5.1), with $\vartheta = \frac{\pi}{2}$, we obtain

$$\bar{\omega}_{\pm} = c\frac{af(r) \pm \sqrt{D}}{(r^2 + a^2) + a^2 f(r)}, \qquad (5.22)$$

with

$$f(r) = \frac{2m}{r} - \frac{B}{2r^3},$$
$$D = r^2 + a^2 - 2mr + \frac{B}{2r}. \qquad (5.23)$$

In [17] it has been shown that for $B \geq B_{\min} = (4/3)^3 m^3$ it holds $D \geq 0$ (see Exercise 4 in Chap. 3), so $\bar{\omega}_{\pm}$ has always two real solutions. It also holds $f(r) \geq 0$ for $r \geq (3B)/(4m)$, that is $r \geq (4/3)m$ for $B = B_{\min}$.

As before, a positive J is chosen and accordingly the a is positive. For positive sign $\bar{\omega}_+ > 0$ and $|\bar{\omega}_+| \geq |\bar{\omega}_-|$, so $\bar{\omega}_+$ describes again the angular frequency of a prograde orbiting photon. Since for large r it holds $D \gg f(r)$, in this range $\bar{\omega}_-$ is negative and corresponds to counter–rotation. For smaller r the term proportional to $1/r$ leads to an increasing $f(r)$, and $\bar{\omega}_-$ might become zero. In classical GR the sphere where $\bar{\omega}_- = 0$ is called the *ergosphere*. Since $g_{33} > 0$, from (5.22) it follows that the radius of the ergosphere is given by the condition $g_{00} = 0$. In pc-GR $-g_{00} > 0$, so ω_- is always negative. However, if $B = B_{\min} + \varepsilon$ with $\varepsilon \ll 1$, at a certain radius $\bar{\omega}_-$ can become very close to zero.

This result of (5.22) is depicted in Fig. 5.3, where the allowed range for circular movement is plotted and pc-GR is compared to Einstein's GR.

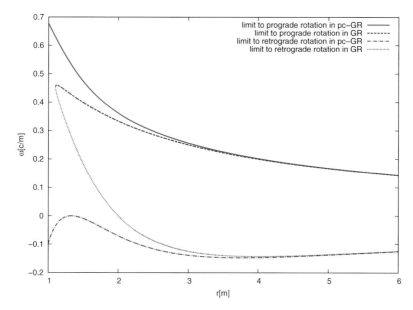

Fig. 5.3 Limits to orbital frequencies of *circular motion* in the equatorial plane. The frequencies of particles moving on *circular orbits* must lie between the shown limiting *curves*. The parameters are $a = 0.995\,m$ and $B = \frac{64}{27}\,m^3$. (Explanation for *gray shades*: The *upper solid curve* corresponds to the limit to prograde rotation in pc-GR and the *lower* one to standard GR. The *upper dashed line* corresponds to the limit to retrograde orbits in standard GR while the *lower* one corresponds to pc-GR)

The curves for pc-GR show a significant different behavior compared to those for Einstein's GR. In standard GR there is a certain radius where the ergosphere begins. For smaller radii, particles have to co-rotate with the central mass [16, 18]. This behavior can be seen in Fig. 5.3 as the limiting curve for counter–rotating orbits changes its sign. The curves finally meet at the event horizon, where all particles have to rotate with the same frequency and in the same direction as the black hole, which is called *frame–dragging*.

The behavior is very different in pc-GR. The curve for counter–rotating orbits also shows the frame-dragging effect, approaching zero from large radial distances to smaller ones. However, approaching the Schwarzschild radius the frame-dragging effects gets weaker and there is no point where the two limiting curves coincide. In fact the point where the curve for counter–rotating orbits reaches zero also marks a maximum of this curve.

This discussion demonstrates that we have definite predictions for the orbital frequency with clear differences to standard GR as already discussed above.

5.1.2 Redshift for Schwarzschild- and Kerr-Type of Solutions

The redshift can be measured through the observation of $Fe - K$ lines, which are observed in accretion discs of galactic black holes, but unfortunately not yet in centers of active galaxies. The new observations of the *Very Large Basis Array* (VLBA) give hope to change that in the near future.

Again we make simplified assumptions, considering the emission of light emitted from a massless particle in its rest-frame. In the section, where a model of the accretion disk is discussed, we will show that corrections have to be implemented due to the motion of the particle. Nevertheless, the main features can be described with the just mentioned simplified assumptions. The light is assumed to be detected by a motionless observer at infinite distance.

The formula for the redshift is obtained relating the proper time of a particle in a strong gravitational field with the coordinate time of the observer, namely

$$ds^2 = c^2 d\tau^2 = g_{00}c^2 dt^2 + g_{11}dr^2 + g_{22}d\vartheta^2 + g_{33}d\varphi^2 + 2g_{03}cdtd\varphi. \quad (5.24)$$

Since the particle is considered at rest the equation simplifies to

$$d\tau^2 = |g_{00}|dt^2. \quad (5.25)$$

In most cases we can assume, that the metric does neither change in the time between two wave peaks (here denoted by τ_0 respectively t_{obs}) nor for the space between the particle and the observer, meanwhile the ray is traveling. With these assumptions, the equation can be integrated and the result is obviously

$$\tau_0 = \sqrt{|g_{00}|}t_{obs}. \quad (5.26)$$

Because the time interval represents the time between two wave peaks, which is the inverse of the frequencies, there are two different time intervals (frequencies), one measured in the rest-frame of the particle and the other in the frame of the observer at infinity. Using (5.26), we obtain

$$\nu_{obs} = \sqrt{|g_{00}|}\nu_0. \quad (5.27)$$

The redshift z is defined as

$$z := \frac{\nu_0 - \nu_{obs}}{\nu_{obs}} = \frac{1}{\sqrt{|g_{00}|}} - 1. \quad (5.28)$$

Using (5.1) we get for the pc-Kerr metric

$$z = \frac{\sqrt{r^2 + a^2 \cos^2(\vartheta)}}{\sqrt{r^2 - 2mr + a^2 \cos^2(\vartheta) + \frac{B}{2r}}} - 1. \quad (5.29)$$

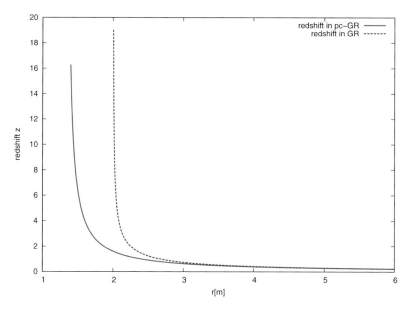

Fig. 5.4 *Redshift* for an emitter in the equatorial plane at the position r in the outside field of a spherically symmetric, uncharged and static mass (Schwarzschild metric) and also for the field at the equator of a rotating mass. B is set to $\frac{64}{27} m^3$

Note that the redshift at the equator of the Kerr solution is exactly the one as in the Schwarzschild case! In Fig. 5.4 the redshift is shown for $\vartheta = \frac{\pi}{2}$, in the limiting case $B = \frac{64}{27} m^3$. A difference between both theories is clearly seen starting between two to three times the Scwarzschild radius. The pc-GR predicts smaller redshifts for the same radius, so that one can look further inside the central mass. In the limiting case for B, the redshift still diverges at $r = \frac{4}{3} m$. Because we required that there is no event-horizon the B has to be at least slightly larger, resulting in very large but finite values for the redshift. Therefore, observing the massive object from far away, looking at events near or in the equatorial plane, the central massive object looks pretty much like a black hole, *though it is not!* Another main feature is that the curve of the redshift looks similar to standard GR but shifted to smaller radial distances.

Again, these features are definite predictions and can be compared to standard GR. In both theories, the redshift is a function in r and from the measured redshift one can deduce the radial distance. As explained in the former section, on the orbital frequency of a particle in a circular orbit, both deduced values of r have to agree for a consistent theory.

This result is not the whole picture for the Kerr solution, since so far we neglected the dependence on the inclination ϑ. Compared to the previous result the other extremes are the poles i.e. $\vartheta = 0$ or $\vartheta = \pi$. This case is shown in Fig. 5.5, where we see a completely different behavior, since the maximal redshift is roughly 1 and the central object should be visible. However, if the central object has an accretion disk (only then our predictions can be verified) a collimated beam is emitted *at the poles*, thus overshining possible emissions from the star's surface.

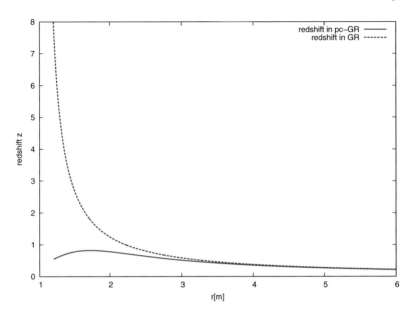

Fig. 5.5 *Redshift* for an emitter at the position r in the outside field of a spherically symmetric, uncharged and rotating mass (Kerr metric) at the poles. The parameter B is again chosen to be $\frac{64}{27} m^3$

Therefore, the observation of a heavy object from strongly different angles could test our predictions. The crucial difference between classical GR black holes and the pc-GR dark stars is that the pc-GR object appear brighter for the same mass. A problem which might arise in this context is that one can hardly distinguish standard black holes from pc-GR objects because one might take a pc-dark star for a standard black hole but underestimate the mass. To prevent this, astronomers should combine various methods to weigh black holes (see e.g. [19] for a review on black hole mass and spin determination).

5.1.3 Effective Potentials and Circular Orbits

As shown in [18] it is very instructive to construct an *effective potential*, which gives information on stable orbits and trajectories a particle will take.

We derive the effective potential for the radial motion of a geodesic in the equatorial plane ($\vartheta = \pi/2$). For this case the variation of the Lagrangian (5.2) leads to the following geodesic equations for t and φ:

$$0 = \frac{d}{ds}\frac{\partial L}{\partial \dot{t}} = \frac{d}{ds}\left(2g_{00}c^2\dot{t} + 2g_{03}c\dot{\varphi}\right),$$
$$0 = \frac{d}{ds}\frac{\partial L}{\partial \dot{\varphi}} = \frac{d}{ds}\left(2g_{33}\dot{\varphi} + 2g_{03}c\dot{t}\right). \tag{5.30}$$

These equations define conserved quantities. The first equation is the energy conservation and the second one the angular momentum conservation. We have

$$g_{00}c\dot{t} + g_{03}\dot{\varphi} =: -\tilde{E},$$
$$g_{33}\dot{\varphi} + g_{03}c\dot{t} =: \tilde{L},$$

(5.31)

where $\tilde{E}$ and $\tilde{L}$ can be identified with the energy $\tilde{E} = E/\mu c^2$ and the angular momentum $\tilde{L} = L/\mu c$ per mass μ of a test particle [16]. An elementary rearrangement yields

$$Dc\dot{t} = g_{03}\tilde{L} + g_{33}\tilde{E},$$
$$D\dot{\varphi} = -g_{03}\tilde{E} - g_{00}\tilde{L},$$

(5.32)

where again $D = \left(-g_{00}g_{33} + g_{03}^2\right)$ was used. Inserting (5.32) in (5.2) yields

$$\dot{r}^2 = \frac{1}{g_{11}D}\left(\tilde{E}^2 g_{33} + 2g_{03}\tilde{L}\tilde{E} + g_{00}\tilde{L}^2 - D\right),$$

(5.33)

which is rewritten as

$$\frac{1}{2}\tilde{E}^2 = \frac{1}{2}\dot{r}^2 + V(r, \tilde{E}, \tilde{L}),$$

(5.34)

with

$$V(r, \tilde{E}, \tilde{L}) = -\frac{1}{2g_{11}D}\left(\tilde{E}^2(g_{33} - Dg_{11}) + 2\tilde{L}\tilde{E}g_{03} + \tilde{L}^2 g_{00} + D\right)$$

$$= \frac{\tilde{L}^2}{2r^2} - \left(\frac{m}{r} - \frac{B}{4r^3}\right)\left(1 + \frac{\left(\tilde{L} + a\tilde{E}\right)^2}{r^2}\right) + \frac{(1 - \tilde{E}^2)a^2}{2r^2} + \frac{1}{2}. \quad (5.35)$$

Exercise 5.4 (Equations (5.32)–(5.35))

Problem. Derive (5.32)–(5.35).

Solution. Let us repeat (5.31):

$$g_{00}c\dot{t} + g_{03}\dot{\varphi} =: -\tilde{E},$$
$$g_{33}\dot{\varphi} + g_{03}c\dot{t} =: \tilde{L}.$$

(5.36)

Subtracting the second equation, divided by g_{33}, from the first equation, divided by g_{03}, we obtain

$$\left(\frac{g_{00}}{g_{03}} - \frac{g_{03}}{g_{33}}\right) c\dot{t} = -\frac{\tilde{E}}{g_{03}} - \frac{\tilde{L}}{g_{33}},$$ (5.37)

from which follows

$$g_{33}\tilde{E} + g_{03}\tilde{L} = -\left[g_{00}g_{33} - g_{03}^2\right] c\dot{t}.$$ (5.38)

With the definition

$$D = g_{03}^2 - g_{00}g_{33},$$ (5.39)

we arrive at

$$Dc\dot{t} = g_{33}\tilde{E} + g_{03}\tilde{L}.$$ (5.40)

The other equation is found by dividing the first equation in (5.36) by g_{00} and subtracting from it the second equation, divided by g_{03}, with the result

$$D\dot{\varphi} = -g_{03}\tilde{E} - g_{00}\tilde{L}.$$ (5.41)

In order to determine $\dot{r}^2$, it is preferable to start from the Lagrangian (5.2). Resolving for $\dot{r}^2$ gives

$$\dot{r}^2 = -\frac{g_{00}}{g_{11}}(c\dot{t})^2 - \frac{g_{33}}{g_{11}}\dot{\varphi}^2 - \frac{2g_{03}}{g_{11}}(c\dot{t})\dot{\varphi} - 1.$$ (5.42)

Substituting (5.40) and (5.41), one gets

$$\dot{r}^2 = \left(-\frac{g_{00}g_{33}^2}{g_{11}D^2} - \frac{g_{33}g_{03}^2}{g_{11}D^2} + 2\frac{g_{33}g_{03}^2}{g_{11}D^2}\right)\tilde{E}^2$$
$$-2\left(\frac{g_{00}g_{33}g_{03}}{g_{11}D^2} - \frac{g_{03}^3}{g_{11}D^2}\right)\tilde{E}\tilde{L}$$
$$+\left(-\frac{g_{00}g_{03}^2}{g_{11}D^2} - \frac{g_{33}g_{00}^2}{g_{11}D^2} + 2\frac{g_{00}g_{03}^2}{g_{11}D^2}\right)\tilde{L}^2$$
$$-1.$$ (5.43)

The factors are treated separately, giving the final expression for $\dot{r}^2$

$$\dot{r}^2 = \frac{1}{g_{11}D}\left[g_{33}\tilde{E}^2 + 2g_{03}\tilde{E}\tilde{L} + g_{00}\tilde{L}^2 - D\right]. \qquad (5.44)$$

In order to get the desired expression, we resolve for $\tilde{E}$ and divide by 2, obtaining

$$\frac{1}{2}\tilde{E} = \frac{1}{2}\dot{r}^2 - \frac{1}{g_{11}D}\left\{\tilde{E}^2\left[g_{33} - g_{11}D\right] + 2g_{03}\tilde{E}\tilde{L} + g_{00}\tilde{L}^2 - D\right\}. \qquad (5.45)$$

The expression in the curly bracket, times its factor, is defined to be $V(r, \tilde{E}, \tilde{L})$.

With (5.34) the radial motion of a geodesic in the equatorial plane is equivalent to the classical motion of a body with unit mass and energy $\tilde{E}^2/2$ in a complicated effective potential $V(r, \tilde{E}, \tilde{L})$. This concept becomes particularly instructive for the Schwarzschild solution. In this case one has $a = 0$, and the effective potential does not depend on $\tilde{E}$:

$$V_S(r, \tilde{L}^2) = \frac{1}{2} - \frac{m}{r} + \frac{\tilde{L}^2}{2r^2} - \frac{m\tilde{L}^2}{r^3} + \frac{B}{4}\left(\frac{1}{r^3} + \frac{\tilde{L}^2}{r^5}\right). \qquad (5.46)$$

The terms $-m/r$ and $\tilde{L}^2/r^2$ correspond to the classical gravitational and centrifugal potential, respectively. In GR the negative term proportional to $1/r^3$ causes the fall of particles into the singularity at $r = 0$, which is avoided in pc-GR due to the repulsive potential proportional to $(1/r^3 + \tilde{L}^2/r^5)$.

Although for the Kerr metric the effective potential is more complicated and depends not only on r and $\tilde{L}$, but also on $\tilde{E}$, it can be used to study the motion of geodesics. Of particular importance are circular orbits, which are given by the simultaneous solutions of $V = \tilde{E}^2/2$ and $\frac{\partial V}{\partial r} = 0$. That is, we consider the set of r-dependent functions $V(r; \tilde{E}, \tilde{L})$ with parameter values $\tilde{E}$ and $\tilde{L}$. These parameters are varied until a curve $V(r; \tilde{E}(r_c), \tilde{L}(r_c))$ is obtained, such that this curve takes on a minimum at $r = r_c$, and has the value $V(r; \tilde{E}(r_c), \tilde{L}(r_c)) = \tilde{E}^2/2$. The radius r_c together with the parameters $\tilde{E}(r_c), \tilde{L}(r_c)$ corresponds to a stable circular orbit. It is convenient to consider the potential $\hat{V} = V - \tilde{E}^2/2$ rather than V. The condition for stable circular orbits is then $\hat{V} = 0$ and $\frac{\partial \hat{V}}{\partial r} = 0$.

Instead of trying to solve these conditions directly via (5.35) and its derivative, we use $\omega \dot{t} = \dot{\varphi}$ as introduced in Sect. 5.1.1 together with (5.32). We get

$$\tilde{E} = -\tilde{t}\frac{cg_{00} + \omega g_{03}}{cg_{03} + \omega g_{33}}. \qquad (5.47)$$

At this point we can use (5.33) for geodesic circular orbits ($\dot{r} = 0$) and insert (5.47):

$$0 = \tilde{L}^2 g_{33} \left(\frac{cg_{00} + \omega g_{03}}{cg_{03} + \omega g_{33}} \right)^2 - \tilde{L}^2 g_{03} \frac{cg_{00} + \omega g_{03}}{cg_{03} + \omega g_{33}} + \tilde{L}^2 g_{00} + D. \qquad (5.48)$$

A straightforward rearrangement of this equation yields (together with (5.47))

$$\tilde{L}^2 = \frac{L^2}{\mu^2 c^2} = \frac{(cg_{03} + \omega g_{33})^2}{-g_{33}\omega^2 - 2g_{03}\omega c - g_{00}c^2},$$

$$\tilde{E}^2 = \frac{E^2}{\mu^2 c^4} = \frac{(cg_{00} + \omega g_{03})^2}{-g_{33}\omega^2 - 2g_{03}\omega c - g_{00}c^2}. \qquad (5.49)$$

In these equations $\tilde{E}$ and $\tilde{L}$ are constants of motion, which correspond to energy and angular momentum of a test particle on a circular geodesic. Both $\tilde{E}$ and $\tilde{L}$ are real numbers, and accordingly the right hand side of (5.49) has to be positive. The denominator corresponds to (5.20) in Sect. 5.1.1 and thus can be written as

$$g_{33}\omega^2 + 2g_{03}\omega c + g_{00}c^2 = g_{33} \left(\omega - \hat{\omega}_+ \right) \left(\omega - \hat{\omega}_- \right), \qquad (5.50)$$

where we have used the respective limiting orbital frequencies for general orbits. It is easy to see, that the constraint of positive $\tilde{E}^2$ and $\tilde{L}^2$ is equivalent to the constraint, that the orbital frequency ω of a stable circular orbits is in the limits given by $\bar{\omega}_\pm$ derived for general orbits.

It will be shown now, that for classical GR the conditions $\hat{V} = 0$ and $\frac{\partial \hat{V}}{\partial r} = 0$ can only be fulfilled for $\tilde{E}^2 < 1$ [20]. Let us, for the moment, consider the classical GR, from (5.35) with $B = 0$ and $\hat{V} = V - \tilde{E}^2/2$ one obtains the expression

$$\hat{V}(r, \tilde{E}, \tilde{L}) = \frac{\tilde{L}^2}{2r^2} - \frac{m}{r} - \frac{m \left(\tilde{L} + a\tilde{E} \right)^2}{r^3} + \frac{(1 - \tilde{E}^2)a^2}{2r^2} + \frac{1}{2} \left(1 - \tilde{E}^2 \right). \qquad (5.51)$$

For $r \to 0$ it holds $\hat{V} \to -\infty$, which has the consequence that if $\hat{V}$ has a minimum at r_c with $\hat{V}(r_c) = 0$, there has to be another root in the interval $(0, r_c)$.

The roots of $\hat{V}$ are identical to the roots of the polynomial

$$P(r, \tilde{E}, \tilde{L}) = r^3 \hat{V}(r, \tilde{E}, \tilde{L})$$
$$= \frac{1}{2} \left(1 - \tilde{E}^2 \right) r^3 - mr^2 + \frac{1}{2} \left(\tilde{L}^2 + \left(1 - \tilde{E}^2 \right) a^2 \right) r$$
$$- m \left(\tilde{L} + a\tilde{E} \right)^2. \qquad (5.52)$$

According to Descartes' rule of signs the number of positive roots is less than or equal to the number of variations in the sign in the polynomial, with multiple roots counted separately [21]. Given a double root at r_c and another root in the interval

$(0, r_c)$, this demands three changes of sign in the polynomial $P(r, \tilde{E}, \tilde{L})$. This is only possible if $\tilde{E}^2 < 1$, so this is a necessary (but not sufficient) condition for the occurrence of a stable circular orbit.

We now turn to the case of pc-GR with $B \neq 0$. From (5.35) with $\hat{V} = V - \tilde{E}^2/2$ we obtain

$$\hat{V}(r, \tilde{E}, \tilde{L}) = \frac{\tilde{L}^2}{2r^2} - \frac{m}{r} - \frac{m\left(\tilde{L} + a\tilde{E}\right)^2}{r^3} + \frac{(1 - \tilde{E}^2)a^2}{2r^2}$$
$$+ \frac{1}{2}\left(1 - \tilde{E}^2\right) + \frac{B}{4r^3} + \frac{B\left(\tilde{L} + a\tilde{E}\right)^2}{4r^5}. \tag{5.53}$$

For $r \to 0$ it holds $\hat{V} \to \infty$ (repulsive potential), whereas for $r \to \infty$ it goes to $\frac{1}{2}\left(1 - \tilde{E}^2\right)$ from below. It follows that if we have a minimum of $\hat{V}$ at r_c with $\hat{V}(r_c) = 0$, for $\tilde{E}^2 \geq 1$ there is another positive root at some value $r' > r_c$. In this case and as well for $\tilde{E}^2 < 1$, there might be additional roots, but their existence is not a necessary condition for a double root which is a minimum at r_c.

The roots of $\hat{V}$ are identical to the roots of the polynomial

$$P(r, \tilde{E}, \tilde{L}) = r^5 \hat{V}(r, \tilde{E}, \tilde{L})$$
$$= \frac{1}{2}\left(1 - \tilde{E}^2\right)r^5 - mr^4 + \frac{1}{2}\left(\tilde{L}^2 + \left(1 - \tilde{E}^2\right)a^2\right)r^3$$
$$+ \left[\frac{B}{4} - m\left(\tilde{L} + a\tilde{E}\right)^2\right]r^2 + B\left(\tilde{L} + a\tilde{E}\right)^2. \tag{5.54}$$

For $\tilde{E}^2 < 1$ there are always two changes of sign in the first three terms, so the necessary condition for the existence of a stable circular orbit is always fulfilled. For $\tilde{E}^2 > 1$, the polynomial has to show three changes of sign (recall that the double root at the minimum is counted separately), so it has to hold

$$\tilde{L}^2 + \left(1 - \tilde{E}^2\right)a^2 > 0,$$
$$\frac{B}{4} - m\left(\tilde{L} + a\tilde{E}\right)^2 < 0. \tag{5.55}$$

It follows, that in pc-GR not the general requirement $\tilde{E}^2 < 1$, but the weaker restriction (5.55) for $\tilde{E}^2 > 1$ is a necessary condition for the occurrence of stable circular orbits.

There might be various combinations r_c, $\tilde{E}(r_c)$, $\tilde{L}(r_c)$ corresponding to stable circular orbits. The last stable orbit can be found by setting $\frac{\partial^2 V}{\partial r^2}|_{r_c} = 0$ Together with (5.49) and (5.35) this is equivalent to

$$g_{33}''(g_{00} + \omega g_{03})^2 + g_{00}''(g_{03} + \omega g_{33})^2 - 2g_{03}''(g_{00} + \omega g_{03})(g_{03} + \omega g_{33})$$
$$+ D''(\omega^2 g_{33} + 2\omega g_{03} + g_{00}) = 0. \tag{5.56}$$

Unfortunately (5.56) has a quite complicated form when we insert the pc-Kerr metric [22]

$$\frac{1}{r\left(4r^5+a^2\left(3B-4mr^2\right)\right)^2}\left[-6a^4\left(15B^2r-32Bmr^3+16m^2r^5\right)\right.$$
$$\pm\,4ar^{5/2}\sqrt{-3B+4mr^2}\left(15B^2+24mr^4(2m+r)-10Br^2(4m+3r)\right)$$
$$+4r^5\left(-15B^2+8mr^4(-6m+r)+2Br^2(20m+3r)\right)$$
$$+a^2\left(45B^3-18B^2r^2(10m+17r)-96mr^6\left(2m^2+5mr+r^2\right)\right.$$
$$\left.+8\,Br^4\left(38m^2+96mr+15r^2\right)\right)\pm 8a^3\left(4mr^{9/2}(4m+3r)\sqrt{-3B+4mr^2}\right.$$
$$\left.\left.-3Br^{5/2}(8m+5r)\sqrt{-3B+4mr^2}+9B^2\sqrt{-3Br+4mr^3}\right)\right]=0\quad(5.57)$$

For $B=0$ this equation greatly simplifies to

$$r^2-6\,mr\pm 8a\sqrt{mr}-3a^2=0,\tag{5.58}$$

which is the same as in [23]. The upper sign here refers to a co-rotating object. Up to now we have not found analytical solutions for (5.57). Numerical investigations showed, that (5.57) has two solutions for retrograde objects (see Fig. 5.6). However only the outer one of these solutions is physically relevant as one has to consider the constraints given by (5.21).

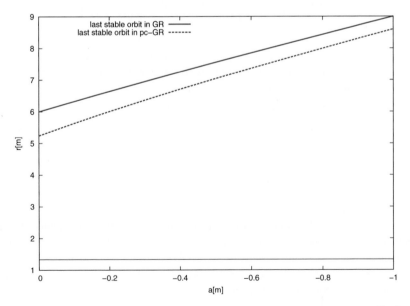

Fig. 5.6 Critical stable orbits for retrograde objects—the parameter B is chosen as $B=\frac{64}{27}m^3$ (For *gray shades*: The *upper* (*solid*) line indicates the last stable orbit in standard GR while the *lower* (*dashed*) curve indicates the last stable orbit in pc-GR)

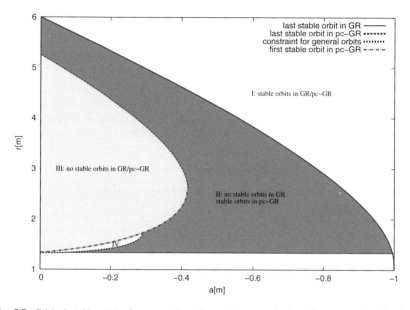

Fig. 5.7 Critical stable orbits for prograde objects (For *gray shades*: The *curves* describe from *top* to *bottom*: *1* The last stable orbit in standard GR (*solid*), *2* the "last" stable orbit in pc-GR (*dashed*), *3* the "first" stable orbit in pc-GR (*dash–dotted*), *4* the limit to general orbits given by (5.21) (*dotted*) and *5* the point where the pc-equations turn imaginary ($r = \frac{4}{3}\,m$, *solid*). In the unshaded area orbits are stable both in GR and pc-GR, whereas in the *dark shaded* area orbits are only stable in pc-GR. Both *lighter shaded* areas do not contain stable orbits at all. The plot is done for a value of $B = \frac{64}{27}\,m^3$

Equation (5.57) also has two solutions for co-rotating objects but only up to a value of $a > 0.416\,m$ (see Fig. 5.7). For higher absolute values of a, Eq. (5.57) has no positive real roots for co-rotating objects. For this case, the second derivative of the potential is always positive, i.e., *a stable orbit always exists*.

As we encounter a new physical phenomenon, namely the existence of two critical orbits compared to only one in the classical Kerr metric, we will investigate Fig. 5.7 a little bit further. All values of $\frac{\partial^2 V}{\partial r^2}$ are negative in the shaded area enclosed by the dark blue dashed (second curve from top) and red (dash–dotted) curve together with the ordinate. Thus in this area circular geodesic orbits are not stable. This is true for both pc-GR and GR. The light shaded area enclosed by the red and black dotted curves is excluded as here the constraint (5.21) is not fulfilled (i.e. $ds^2 < 0$ in this area). But on the outside—especially for values of $a > 0.416\,m$—we have $\frac{\partial^2 V}{\partial r^2} > 0$ (and $E \leq \mu c^2$) which means that the circular orbits are stable again for radii greater than $\frac{4}{3}\,m$. Therefore we call the dark blue (dashed) curve the "last" stable orbit with quotation marks as there is another curve (red dash–dotted) which we call the "first" stable orbit from which on orbits are stable again. This is a significant difference to standard GR.

For smaller radii than $\frac{4}{3} m$ the equations turn imaginary and the orbital frequency is imaginary. However, $r = \frac{4}{3} m$ is the position of the global minimum of g_{00} (and therefore the effective potential), describing the final radius of a large mass. In other words, it has no sense to go further below, due to the presence of mass. For a description at lower radial distances, the distribution of the star's mass has to be included.

In total the physical behavior for retrograde orbits is not that different to the standard GR except that the last stable orbit is a little bit closer to the central mass. For prograde orbits however we get a new physical behavior.

5.1.4 Galactic "Black-Holes" and Observational Verification by pc-GR

As mentioned in the introduction to this chapter, observations of so-called "galactic black holes" seem to contradict standard GR [6–10]. The correctness of the conclusions in this section depends on the interpretation, that we can assume the *Quasi Periodic Oscillations* (QPO) observed to be interpreted as local emissions in the accretion disk, following its motion. This interpretation is correct for the accretion disk in the center of active galaxies, because there are no nearby stellar partners. By extrapolation, we assume the same interpretation for the galactic black holes.

And there is the problem (not for pc-GR but for standard GR): *the measured orbital frequency give radii of* 20–50 r_g *(the standard definition of r_g is just m, i.e. half of the Schwarzschild radius), while from the measured redshift radii of* 2–5 r_g *are deduced* [6–10] (see also [24]). This is a big difference!

As an example we discuss the galactic object, called GRO J1655-40. Its estimated mass is $M = 0.30 \pm 0.5$ solar masses and the spin parameter is $a = 0.92 \pm 0.2$. In Fig. 5.8 the frequency of the observed QPO's is plotted versus the radial distance. Two curves are visible, the one with a steady increase is due to GR and the one with the maximum is pc-GR. The thickness of the curves is due to observational errors. The range of possible radial distances, as deduced from the frequency measurements, is indicated by the two vertical dashed lines on the right part of the figure. The horizontal dashed lines indicate the measured limits of the frequencies. From there, the r should be between 17 and 14 r_g. The redshift, however, gives values around 2 r_g, as indicated by the vertical dashed lines on the left part of the figure, as deduced from the redshift curves of GR. Note, that in pc-GR the redshift curve is shifted to the left, which brings it into agreement with pc-GR. Also it is in agreement with the measured orbital frequency!

Thus, pc-GR permits to have large redshifts with low orbital frequency at small radial distances! This presents the first evidence that pc-GR describes the physics correctly near a large massive object, with a very strong gravitational field, which standard GR can not do.

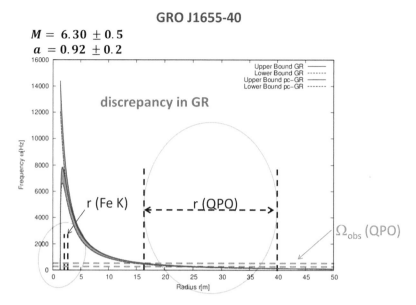

Fig. 5.8 Angular orbital frequency ω, in units of $\frac{c}{m}$, versus the radial distance, in units of $r_g = m$. The steady increasing *curve* toward smaller r (*blue curve*) is the result of GR, while the other, with the maximum, is pc-GR. The width corresponds to the insecurity in knowing the mass and the spin of the central object

GR J1655-40 is not the only observed galactic "black hole", in contradiction with standard GR. There are three further objects [6–10], with the names GX 339-4, J1752-223 and XTE J1550-564.

There are more possible observations in near future, which may provide further evidence in favor of pc-GR. This is, picturing the appearance of an accretion disk as seen from far away. This will be the topic of the next section, related to the *raytracing method*.

5.2 Raytracing Method and the Imaging of the Accretion Disk

A so-called black hole can not be observed when completely isolated, because it is black. In pc-GR this object is rather a *dark star*, but it behaves very similar to a black hole. Only when mass is falling in, like through an accretion disk, the light emitted from this infalling matter can be observed, and here GR and pc-GR predict different behavior! In the last section we already demonstrated that particles move differently in circular orbits. In this section the accretion disk, if formed, is investigated and we will see that also GR and pc-GR will give distinct predictions, with clearer observable consequences. We will *image* this central object and its strong gravitational field

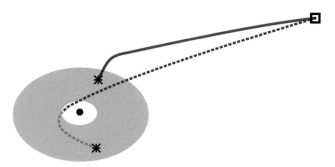

Fig. 5.9 Illustration of the raytracing technique: Two rays, originating from the accretion disk, are shown. The *red line* represents a light path which reaches on a geodesic path the observer at infinity, having been distorted by the gravitational field. The *blue-dotted light curve* represents a second order effect, where the light makes a near complete turn around the star. The raytracing method follows the ray back, starting from the observer. In each path the conservation laws and the Carter constant are verified, until one reaches a point at the accretion disk. In such a way, only light rays are taken into account which really reaches the observer, which reduces the numerical effort enormously

by following light rays coming from a source near this object. A powerful standard technique is called *raytracing*. The basic idea is to follow light rays (on null geodesics) in a curved background space-time from their point of emission, for example from a point of an accretion disk, around a massive object (Technically this is not correct. It is computationally less expensive to follow light rays from an observers screen back to their point of emission.). This permits us to create an image in the direct neighborhood of an massive object. The raytracing method is illustrated in Fig. 5.9. There are numerous groups using raytracing for this purpose [14, 15, 25, 26]. Aside from creating an image, it is also possible to calculate emission line profiles using the same technique, but adding in a second step the evaluation of an integral for the spectral flux. This is of particular interest as the emission profile of the iron Kα line is one of the few good observables in regions with strong gravitation, from which the redshift can be deduced (see last section). (For more details, see [27]).

In the following, the theoretical background on the methods used will briefly reviewed, where we will also discuss the model to describe accretion disks. After that we will present results obtained with the open source ray-tracing code GYOTO [26] for the simulation of disk images and emission line profiles.

5.2.1 Theoretical Background

For convenience, the velocity of light c is *set to one* in order to agree with the standard notation in models for the accretion disk.

We will use the Boyer-Lindquist coordinates of the Kerr metric and its pseudo-complex equivalent, which is written in a slightly different form than in [17] as (see also Exercise 5.5)

$$g_{00} = -\left(1 - \frac{\psi}{\Sigma}\right),$$

$$g_{11} = \frac{\Sigma}{\Delta},$$

$$g_{22} = \Sigma,$$

$$g_{33} = \left((r^2 + a^2) + \frac{a^2\psi}{\Sigma}\sin^2\vartheta\right)\sin^2\vartheta,$$

$$g_{03} = -a\frac{\psi}{\Sigma}\sin^2\vartheta, \tag{5.59}$$

with

$$\Sigma = r^2 + a^2\cos^2\vartheta,$$

$$\Delta = r^2 + a^2 - \psi,$$

$$\psi = 2\,mr - \frac{B}{2r}. \tag{5.60}$$

The $m = \kappa M$ is the gravitational radius r_g of a massive object. One can easily see that this metric differs from the standard Kerr metric only in the use of the function ψ which reduces to $2\,mr$ in the limit $B = 0$. Bearing this in mind one can simply follow the derivation of the Lagrange equations given, as in [28], which are the basis for the implementation in [26]) and modify the occurrences of the Boyer-Lindquist Δ-function and the new introduced ψ.

Exercise 5.5 (Equation 5.59)

Problem. Relate (5.59) with (5.1).

Solution. Substituting the expressions (5.60) in (5.59), leads to

$$g_{00} = -\left[1 - \frac{\psi}{\Sigma}\right] = -\left[1 - \frac{\left(2\,mr - \frac{B}{2r}\right)}{r^2 + a^2\cos\vartheta}\right]$$

$$= -\frac{\left[r^2 + a^2\cos^2\vartheta - 2\,mr + \frac{B}{2r}\right]}{r^2 + a^2\cos^2\vartheta},$$

$$g_{11} = +\frac{\Sigma}{\Delta} = \frac{r^2 + a^2\cos^2\vartheta}{r^2 + a^2 - 2\,mr + \frac{B}{2r}},$$

$$g_{22} = +\Sigma = r^2 + a^2\cos^2\vartheta,$$

$$g_{33} = \left[(r^2 + a^2) + \frac{a^2\psi}{\Sigma}\sin^2\vartheta\right]$$

$$= (r^2 + a^2)\sin^2\vartheta + \sin^4\vartheta\frac{a^2\left(2\,mr - \frac{B}{2r}\right)}{r^2 + a^2\sin^2\vartheta}. \tag{5.61}$$

This is the expression for the Kerr metric as given in (5.1) at the start of this chapter.

In oder to derive the final equations of motion all the possible conserved quantities along geodesics are introduced, which are:

(i) The test particle's mass m_0,
(ii) the energy at infinity E,
(iii) the angular momentum L_z and
(iv) the Carter constant Q [28, 29] (see also explanations further below).

The usual way to proceed from this, is to use the Jacobi formalism [29] and demand separability of Hamilton's principal function (see also Exercise (5.6) further below).

$$S = -\frac{1}{2}\lambda - Et + L_z\varphi + S_\vartheta + S_r. \tag{5.62}$$

The λ is an affine parameter and S_r and S_ϑ are functions in r and ϑ, respectively. Demanding separability for (5.62) leads [29] to

$$\left(\frac{dS_r}{dr}\right)^2 = \frac{R}{\Delta^2}, \left(\frac{dS_\vartheta}{d\vartheta}\right)^2 = \Theta, \tag{5.63}$$

with the auxiliary functions

$$R(r) := \left[(r^2 + a^2)E - aL_z\right]^2 \\ - \Delta\left[Q + (aE - L_z)^2 + m_0^2 r^2\right],$$
$$\Theta(\vartheta) := Q - \left[\frac{L_z^2}{\sin^2\vartheta} - a^2 E^2 + m_0^2 a^2\right]\cos^2\vartheta. \tag{5.64}$$

The second equation contains the Carter constant Q.

Taking these together with

$$\dot{x}^\mu = g^{\mu\alpha} p_\alpha = g^{\mu\alpha}\frac{\partial S}{\partial x^\alpha} \tag{5.65}$$

leads to a set of first order equations of motion in the coordinates

$$\dot{t} = \frac{1}{\Sigma\Delta}\left\{\left[(r^2 + a^2)^2 + a^2\Delta\sin^2\vartheta\right]E - a\psi L_z\right\},$$
$$\dot{r} = \pm\frac{\sqrt{R}}{\Sigma},$$
$$\dot{\vartheta} = \pm\frac{\sqrt{\Theta}}{\Sigma},$$
$$\dot{\varphi} = \frac{1}{\Sigma\Delta}\left[\left(\frac{\Delta}{\sin^2\vartheta} - a^2\right)L_z + a\psi E\right], \tag{5.66}$$

where the dot represents the derivative with respect to the proper time τ.

In [28] it is however argued, that those equations contain an ambiguity in the sign for the radial and azimuthal velocities. In addition to that, using Hamilton's principle to get the geodesics leads to the integral equation [29]

$$\int_{r_{em}}^{r_{obs}} \frac{dr}{\sqrt{R}} = \int_{\vartheta_{em}}^{\vartheta_{obs}} \frac{d\vartheta}{\sqrt{\Theta}}.$$

(5.67)

To solve this equation, in [14, 15] one makes use of the fact that R is a fourth order polynomial in r. In pc-GR, however, the order of the polynomial is larger, due to the additional term, thus the procedure used in [14, 15] can not be followed anymore.

Fortunately, the equations used in GYOTO are based on the use of different equations of motion derived in [28], which does not suffer from the sign ambiguity. Therefore we follow [28] who make use of the Hamiltonian formulation in addition to the separability of Hamilton's principal function: The canonical 4-momentum of a particle is given as

$$p^\mu := \dot{x}^\mu.$$

(5.68)

The Lagrangian is given by

$$L = \frac{1}{2} g_{\mu\nu} \dot{x}^\mu \dot{x}^\nu.$$

(5.69)

Given explicitly in their covariant form the momenta one has [28]

$$p_0 = -\left(1 - \frac{\psi}{\Sigma}\right) \dot{t} - \frac{\psi a \sin^2 \vartheta}{\Sigma} \dot{\varphi},$$

$$p_1 = \frac{\Sigma}{\Delta} \dot{r},$$

$$p_2 = \Sigma \dot{\vartheta},$$

$$p_3 = \sin^2 \vartheta \left(r^2 + a^2 + \frac{\psi a^2 \sin^2 \vartheta}{\Sigma}\right) \dot{\varphi} - \frac{\psi a \sin^2 \vartheta}{\Sigma} \dot{t}.$$

(5.70)

After a short calculation the Hamiltonian $H = p_\mu \dot{x}^\mu - L = \frac{1}{2} g^{\mu\nu} p_\mu p_\nu$ can be rewritten as [28] (see also Exercise 5.6)

$$H = \frac{\Delta}{2\Sigma} p_1^2 + \frac{1}{2\Sigma} p_2^2 - \frac{R(r) + \Delta\Theta(\vartheta)}{2\Delta\Sigma} - \frac{m_0}{2}.$$

(5.71)

The momenta associated with time t and azimuth φ are conserved and are identified with the energy at infinity $p_0 = -E$ and the angular momentum $p_3 = L_z$, respectively [29].

Hamilton's equations $\dot{x}_\mu = \frac{\partial H}{\partial p_\mu}$ and $\dot{p}_\mu = -\frac{\partial H}{\partial x_\mu}$ yield the wanted equations of motion [26, 28], replacing the p_0 and p_3 by the constants of motion $-E$ and L_z:

$$\dot{t} = \frac{1}{2\Delta\Sigma} \frac{\partial}{\partial E}(R + \Delta\Theta),$$

$$\dot{r} = \frac{\Delta}{\Sigma} p_1,$$

$$\dot{\vartheta} = \frac{1}{\Sigma} p_2,$$

$$\dot{\varphi} = -\frac{1}{2\Delta\Sigma} \frac{\partial}{\partial L_z}(R + \Delta\Theta),$$

$$\dot{p}_0 = 0,$$

$$\dot{p}_1 = -\left(\frac{\Delta}{2\Sigma}\right)_{|r} p_1^2 - \left(\frac{1}{2\Sigma}\right)_{|r} p_2^2 + \left(\frac{R + \Delta\Theta}{2\Delta\Sigma}\right)_{|r},$$

$$\dot{p}_2 = -\left(\frac{\Delta}{2\Sigma}\right)_{|\vartheta} p_1^2 - \left(\frac{1}{2\Sigma}\right)_{|\vartheta} p_2^2 + \left(\frac{R + \Delta\Theta}{2\Delta\Sigma}\right)_{|\vartheta},$$

$$\dot{p}_3 = 0. \tag{5.72}$$

The notation $_{|r}$ and $_{|\vartheta}$ stand for the partial derivatives with respect to r and ϑ.

Exercise 5.6 (Hamilton-Jacobi formalism)

Problem. Give a detailed description of the Hamilton-Jacobi formalism used.

Solution. The basis for the raytracing technique are the geodesic equations for orbits. We part from the Lagrange function

$$L = \frac{1}{2} g_{\mu\nu} \dot{x}^\mu \dot{x}^\nu. \tag{5.73}$$

With the new signature used, one has $g_{\mu\nu} \dot{x}^\mu \dot{x}^\nu = -\zeta$, where $\zeta = 1$ for time-like trajectories, $\zeta = -1$ for space-like and $\zeta = 0$ for null geodesics. The corresponding Hamilton function is given by

$$H = \frac{1}{2} g^{\mu\nu} p_\mu p_\nu = -\frac{1}{2}\zeta \tag{5.74}$$

with the momenta

$$p_\mu = g_{\mu\nu} \dot{x}^\nu. \tag{5.75}$$

As the metric, we use the pseudo-complex equivalent to the Kerr solution, slightly deviated from the standard definition. This we listed above and here, for convenience, it will be repeated namely

$$g_{00} = -\left(1 - \frac{\psi}{\Sigma}\right),$$

$$g_{11} = \frac{\Sigma}{\Delta},$$

$$g_{22} = \Sigma,$$

$$g_{33} = \left((r^2 + a^2) + \frac{a^2\psi}{\Sigma}\sin^2\vartheta\right)\sin^2\vartheta,$$

$$g_{03} = -a\frac{\psi}{\Sigma}\sin^2\vartheta. \tag{5.76}$$

We did introduce besides the known auxiliary functions Σ and Δ the additional auxiliary function ψ, i.e.,

$$\Sigma = r^2 + a^2\cos^2\vartheta,$$

$$\Delta = r^2 + a^2 - \psi,$$

$$\psi = 2mr - \frac{B}{2r}. \tag{5.77}$$

The components of the inverse metric are given by

$$g^{00} = -\frac{(r^2 + a^2)^2 - a^2\Delta\sin^2\vartheta}{\Sigma\Delta},$$

$$g^{11} = \frac{\Delta}{\Sigma},$$

$$g^{22} = \frac{1}{\Sigma},$$

$$g^{33} = \frac{\Delta - a^2\sin^2\vartheta}{\Sigma\Delta\sin^2\vartheta},$$

$$g^{03} = -\frac{a\psi}{\Sigma\Delta}. \tag{5.78}$$

With (5.76) the components of the momentum are given by

$$p_t = g_{00}\dot{t} + g_{03}\dot{\varphi} = -\left(1 - \frac{\psi}{\Sigma}\right)\dot{t} - a\frac{\psi}{\Sigma}\sin^2\vartheta\dot{\varphi},$$

$$p_r = g_{11}\dot{r} = \frac{\Sigma}{\Delta}\dot{r},$$

$$p_\vartheta = g_{22}\dot{\vartheta} = \Sigma\dot{\vartheta},$$

$$p_\varphi = g_{30}\dot{t} + g_{33}\dot{\varphi} = -a\frac{\psi}{\Delta}\sin^2\vartheta\dot{t} + \left[(r^2 + a^2) + a^2\frac{\psi}{\Sigma}\sin^2\vartheta\right]\sin^2\vartheta\dot{\varphi}. \tag{5.79}$$

The symmetry of space-time, which we describe by the present metric, determines certain restrictions, i.e., the momentum components $p_t = -E$, the energy, and $p_\varphi = L_z$, the angular momentum, both at infinity. In addition, the condition of a constant Hamilton function corresponds to the conservation of the rest-mass of the particle. A forth constant of motion can be found through the requirement of separability of the Hamilton-Jacobi equation (this leads to the Carter constant). The Hamilton-Jacobi equation is given by

$$\frac{\partial S}{\partial \lambda} = \frac{1}{2} g^{\alpha\beta} \left(\frac{\partial S}{\partial x^\alpha} \frac{\partial S}{\partial x^\beta} \right), \tag{5.80}$$

and with the requirement of separability one can apply the following ansatz

$$S = -\frac{1}{2}\zeta\lambda - Et + L_z\varphi + S_\vartheta + S_r, \tag{5.81}$$

where λ is an affine parameter and where the two functions S_ϑ and S_r depend only on ϑ and r, respectively. Substituting this ansatz into the Hamilton-Jacobi equation yields

$$2\frac{\partial S}{\partial \lambda} = g^{00}\left(\frac{\partial S}{\partial t}\right)^2 + g^{11}\left(\frac{\partial S}{\partial r}\right)^2 + g^{22}\left(\frac{\partial S}{\partial \vartheta}\right)^2 + g^{33}\left(\frac{\partial S}{\partial \varphi}\right)^2 + 2g^{03}\frac{\partial S}{\partial t}\frac{\partial S}{\partial \varphi}$$

$$\Leftrightarrow -\zeta = g^{00}E^2 + g^{33}L_z^2 - 2g^{03}EL_z + g^{11}\left(\frac{\partial S}{\partial r}\right)^2 + g^{22}\left(\frac{\partial S}{\partial \vartheta}\right)^2. \tag{5.82}$$

Now, we substitute the inverse matrix elements of the Kerr metric (5.78) and with some minor manipulations we obtain

$$-\zeta = -\frac{(r^2 + a^2)^2 - a^2\Delta\sin^2\vartheta}{\Sigma\Delta}E^2 + \frac{\Delta}{\Sigma}\left(\frac{\partial S}{\partial r}\right)^2 + \frac{1}{\Sigma}\left(\frac{\partial S}{\partial \vartheta}\right)^2$$

$$+ \frac{\Delta - a^2\sin^2\vartheta}{\Sigma\Delta\sin^2\vartheta}L_z^2 + 2\frac{a\psi}{\Sigma\Delta}EL_z$$

$$\Leftrightarrow -\zeta\Sigma = \Delta\left(\frac{\partial S}{\partial r}\right)^2 + \left(\frac{\partial S}{\partial \vartheta}\right)^2$$

$$- \frac{(r^2 + a^2)^2}{\Delta}E^2 + a^2\sin^2\vartheta E^2 + \frac{1}{\sin^2\vartheta}L_z^2 - \frac{a^2}{\Delta}L_z^2 + \frac{2a\psi}{\Delta}EL_z$$

$$\Leftrightarrow -\zeta\Sigma = \Delta\left(\frac{\partial S}{\partial r}\right)^2 + \left(\frac{\partial S}{\partial \vartheta}\right)^2$$

$$- \frac{1}{\Delta}\left((r^2 + a^2)E - aL_z\right)^2 - \frac{2}{\Delta}(r^2 + a^2)aEL_z$$

$$+ \frac{1}{\sin^2\vartheta}\left(-L_z + a\sin^2\vartheta E\right)^2 + 2aEL_z + \frac{2a\psi}{\Delta}EL_z. \tag{5.83}$$

All mixing terms vanish, i.e.,

$$-\frac{2}{\Delta}(r^2+a^2)aEL_z + 2aEL_z + \frac{2a\psi}{\Delta}EL_z = 2EL_z\left(-\frac{1}{\Delta}\left((r^2+a^2)a - a\psi\right)+a\right)$$
$$= 2EL_z\,(a-a) = 0, \qquad (5.84)$$

Therefore, we get

$$-\zeta\Sigma = \Delta\left(\frac{dS_r}{dr}\right)^2 + \left(\frac{dS_\vartheta}{d\vartheta}\right)^2$$
$$-\frac{1}{\Delta}\left((r^2+a^2)E - aL_z\right)^2 + \frac{1}{\sin^2\vartheta}\left(-L_z + a\sin^2\vartheta E\right)^2. \quad (5.85)$$

One further algebraic manipulations leads us to the separation of the equation. The following relation is valid:

$$\frac{1}{\sin^2\vartheta}\left(-L_z + a\sin^2\vartheta E\right)^2 = \frac{L_z^2}{\sin^2\vartheta} + a^2\sin^2\vartheta E^2 - 2aL_zE$$
$$= \frac{L_z^2}{\sin^2\vartheta} + a^2E^2 - a^2\cos^2\vartheta E^2 - 2aL_zE + L_z^2 - L_z^2$$
$$= \frac{L_z^2 - L_z^2\sin^2\vartheta}{\sin^2\vartheta} - a^2\cos^2\vartheta E^2 + (aE + L_z)^2$$
$$= \left(\frac{L_z^2}{\sin^2\vartheta} - a^2E^2\right)\cos^2\vartheta + (aE - L_z)^2,$$

which we can substitute into (5.85) and finally makes it possible to separate the equation into one depending on ϑ only and the other in r:

$$-\zeta r^2 - \Delta\left(\frac{dS_r}{dr}\right)^2 + \frac{1}{\Delta}\left((r^2+a^2)E - aL_z\right)^2 - (aE - L_z)^2$$
$$= \left(\zeta a^2 + \frac{L_z^2}{\sin^2\vartheta} - a^2E^2\right)\cos^2\vartheta + \left(\frac{dS_\vartheta}{d\vartheta}\right)^2. \qquad (5.86)$$

Both sides depend on different variables, which implies that both sides can be set to a constant Q, *which is called the Carter constant.* Therefore, we can introduce the functions $R(r)$ and $\Theta(\vartheta)$ and write

$$\left(\frac{dS_r}{dr}\right)^2 = \frac{R}{\Delta^2} \text{ and, } \left(\frac{dS_\vartheta}{d\vartheta}\right)^2 = \Theta, \qquad (5.87)$$

where the two introduced functions are given by

$$R := \left[(r^2 + a^2)E - aL_z \right]^2 - \Delta \left[\mathscr{C} + (aE - L_z)^2 + \zeta r^2 \right],$$

$$\Theta := Q - \left(a^2(\zeta - E^2) + \frac{L_z^2}{\sin^2 \vartheta} \right) \cos^2 \vartheta. \tag{5.88}$$

This allows us to built the equations of motion. For this, we use

$$\dot{x}^\mu = g^{\mu\alpha} p_\alpha = g^{\mu\alpha} \frac{\partial S}{\partial x^\alpha} \tag{5.89}$$

and obtain

$$\dot{x}^0 = \dot{t} = \frac{1}{\Sigma \Delta} \left\{ \left[(r^2 + a^2)^2 + a^2 \Delta \sin^2 \vartheta \right] E - a\psi L_z \right\},$$

$$\dot{x}^1 = \dot{r} = \pm \frac{\sqrt{R}}{\Sigma},$$

$$\dot{x}^2 = \dot{\vartheta} = \pm \frac{\sqrt{\Theta}}{\Sigma},$$

$$\dot{x}^3 = \dot{\varphi} = \frac{1}{\Sigma \Delta} \left[\left(\frac{\Delta}{\sin^2 \vartheta} - a^2 \right) L_z + a\psi E \right]. \tag{5.90}$$

However, we will choose a different path than the standard one, in order to be in line with [26], the program which uses the raytracing method.

We write (5.74) explicitly, inserting the elements of the inverse metric, i.e.,

$$H = \frac{1}{2} \frac{\Delta}{\Sigma} p_r^2 + \frac{1}{2} \frac{1}{\Sigma} p_\vartheta^2 + f(r, \vartheta, p_t, p_\varphi). \tag{5.91}$$

The remaining terms were joined into the auxiliary function $f(r, \vartheta, p_t, p_\varphi)$. The dependence of f on r and ϑ is introduced via $g^{\mu\nu}$. Now, we substitute $p_r = \frac{\partial S}{\partial r} = \frac{dS_r}{dr}$ and $p_\vartheta = \frac{\partial S}{\partial \vartheta} = \frac{dS}{d\vartheta}$, with the help of (5.87) and (5.88)

$$H = \frac{R}{2\Delta\Sigma} + \frac{\Theta}{2\Sigma} + f(r, \vartheta, p_t, p_\varphi). \tag{5.92}$$

Here, we can use again (5.74) and determine with it the function f, namely

$$H = \frac{R}{2\Delta\Sigma} + \frac{\Theta}{2\Sigma} + f(r, \vartheta, p_t, p_\varphi) = -\frac{1}{2}\zeta$$

$$\Rightarrow f(r, \vartheta, p_t, p_\varphi) = -\frac{R + \Delta\Theta}{2\Delta\Sigma} - \frac{1}{2}\zeta. \tag{5.93}$$

With this, the Hamilton function acquires the form

$$H = \frac{\Delta}{2\Sigma} p_r^2 + \frac{1}{2\Sigma} p_\vartheta^2 - \frac{R + \Delta\Theta}{2\Delta\Sigma} - \frac{1}{2}\zeta. \tag{5.94}$$

The dependence on p_t and p_φ is implicitly contained in the functions R and Θ, due to their dependence on $-E$ and L_z.

The equations of motion are obtained, using the Hamilton equations, i.e.,

$$\dot{p}_\mu = -\frac{\partial H}{\partial x^\mu} \, , \, \dot{x}_\mu = \frac{\partial H}{\partial p_\mu}. \tag{5.95}$$

The results is

$$\dot{t} = \frac{\partial H}{\partial p_t} = -\frac{\partial H}{\partial E} = \frac{1}{2\Delta\Sigma}\frac{\partial}{\partial E}(R + \Delta\Theta),$$

$$\dot{r} = \frac{\partial H}{\partial p_r} = \frac{\Delta}{\Sigma} p_r,$$

$$\dot{\vartheta} = \frac{\partial H}{\partial p_\vartheta} = \frac{1}{\Sigma} p_\vartheta,$$

$$\dot{\varphi} = \frac{\partial H}{\partial p_\varphi} = \frac{\partial H}{\partial L_z} = -\frac{1}{2\Delta\Sigma}\frac{\partial}{\partial L_z}(R + \Delta\Theta),$$

$$\dot{p}_t = -\frac{\partial H}{\partial t} = 0,$$

$$\dot{p}_r = -\frac{\partial H}{\partial r} = -\left(\frac{\Delta}{2\Sigma}\right)_{|r} p_r^2 - \left(\frac{1}{2\Sigma}\right)_{|r} p_\vartheta^2 + \left(\frac{R + \Delta\Theta}{2\Delta\Sigma}\right)_{|r},$$

$$\dot{p}_\vartheta = -\frac{\partial H}{\partial \vartheta} = -\left(\frac{\Delta}{2\Sigma}\right)_{|\vartheta} p_r^2 - \left(\frac{1}{2\Sigma}\right)_{|\vartheta} p_\vartheta^2 + \left(\frac{R + \Delta\Theta}{2\Delta\Sigma}\right)_{|\vartheta},$$

$$\dot{p}_\varphi = -\frac{\partial H}{\partial \varphi} = 0. \tag{5.96}$$

The $_{|r}$ and $_{|\vartheta}$ denote the partial derivatives through r and ϑ respectively.

Equation (5.96) represents advantages compared to the classical used form, which are not unique in the sign of the radial and polar coordinate. Therefore, in the classical case one has to discuss two possibilities.

In obtaining the geodesic equations, the use of the Hamilton principle yields the integral equation

$$\int_{r_{em}}^{r_{obs}} \frac{dr}{\sqrt{R}} = \int_{\vartheta_{em}}^{\vartheta_{obs}} \frac{d\vartheta}{\sqrt{\Theta}}. \tag{5.97}$$

In order to solve this equation, one uses the fact that R is a polynomial of fourth order in r. However, in the modified metric in pc-GR this is not the case anymore. As a result, one can not use the solution proposed in literature.

The Eq. (5.96) do not contain square roots anymore and, thus, can be solved without having to consider two distinct cases. They can be applied directly to the modified Kerr metric.

In addition to the modification of the metric and thus the evolution equations one has to modify the orbital frequency of particles around a compact object. This has been done in the previous section [30] and the formula for the orbital frequencies is repeated:

$$\omega_\pm = \frac{1}{a \pm \sqrt{\frac{2r}{h(r)}}}, \tag{5.98}$$

where ω_+ describes prograde motion and $h(r) = \frac{2m}{r^2} - \frac{3B}{2r^4}$. Equation (5.98) reduces to the well known $\omega_\pm = \frac{1}{a \pm \sqrt{\frac{r^3}{m}}}$ for $B = 0$.

Finally the concept of an *Innermost Stable Circular Orbit* (ISCO) has to be revised, as the pc-equivalent of the Kerr metric only shows an ISCO for some values of the spin parameter a. For values of a greater than $0.416\,m$ and $B = \frac{64}{27}\,m^3$ there is no region of unstable orbits anymore [30] (see also previous section). We can not explain all the details of the disk model, because this would be out of the scope of this section. We refer to the literature.

After including all those changes due to correction terms of the pseudo-complex equivalent of the Kerr metric one can straightforwardly adapt the calculations done in GYOTO. The modified GYOTO program, which includes the B term, can be found in [31].

5.2.1.1 Deducing the Observed Flux

Here, the physical observables, used in raytracing studies, are discussed. We will restrict to *null-geodesics*, i.e. light rays. The first quantity of interest is the intensity of the radiation, emitted between a point s_0 and the position s in the emitters frame, is given by [26, 32]

$$I_\nu(s) = \int_{s_0}^{s} \exp\left(-\int_{s'}^{s} \alpha_\nu(s'')ds''\right) j_\nu(s')ds'. \tag{5.99}$$

Here α_ν is the absorption coefficient and j_ν the emission coefficient in the comoving frame.

Using the invariant intensity $\mathscr{I} = I_\nu/\nu^3$ [18], the observed intensity is

$$I_{\nu_{obs}} = g^3 I_{\nu_{em}}, \tag{5.100}$$

where the relativistic generalized redshift factor $g := \frac{\nu_{obs}}{\nu_{em}}$ was introduced. The flux F is the quantity observed

$$\mathrm{d}F_{\nu_{obs}} = I_{\nu_{obs}} \cos \vartheta \, \mathrm{d}\Omega, \tag{5.101}$$

where ϑ describes the angle between the normal of the observers screen and the direction of incidence and Ω gives the solid angle in which the observer sees the light source [26].

This emission model considers a geometrically thin, infinite accretion disk, proposed by Page and Thorne [33]. The intensity profile is strongly dependent on the metric used and some modifications have to be done. Fortunately most of the results of [33] can be inherited and only at the end one has to insert the modified metric. Equation (12) in [33]

$$f = -\omega_{|r}(E - \omega L_z)^{-2} \int_{r_{ms}}^{r} (E - \omega L_z) L_{z|r} \, \mathrm{d}r \tag{5.102}$$

builds the core for the computation of the flux [33]

$$F = \frac{\dot{M}_0}{4\pi \sqrt{-g}} f. \tag{5.103}$$

$\dot{M}_0$ is the mass flux through the disk. Assuming $\dot{M}_0 = 1$ as done automatically in the program published in [26], just to define the units, and observing that the square root of the determinant of the metric $\sqrt{-g}$ is the same for both GR and pc-GR, we see that the only difference in the flux lies in the function f given by (5.102).

Exercise 5.7 (Determinant of the metric in GR and pc-GR)

Problem. Show that for the Kerr metric $\sqrt{-g}$ is the same in GR and pc-GR.

Solution. The metric in matrix form is given by

$$(g_{\mu\nu}) = \begin{Bmatrix} -\left(1 - \frac{\psi}{\Sigma}\right) & 0 & 0 & -a\frac{\psi}{\Sigma}\sin^2\vartheta \\ 0 & \frac{\Sigma}{\Delta} & 0 & 0 \\ 0 & 0 & \Sigma & 0 \\ -a\frac{\psi}{\Sigma}\sin^2\vartheta & 0 & 0 & (r^2 + a^2)\sin^2\vartheta + a^2\frac{\psi}{\Sigma}\sin^4\vartheta \end{Bmatrix}. \tag{5.104}$$

The only dependence on the parameter B is in ψ and Δ, which depends on ψ.

With this, the negative determinant of the metric, called $-g$, is given by

$$
\begin{aligned}
-g &= -\frac{\Sigma^2}{\Delta}\left\{-\left(1-\frac{\psi}{\Sigma}\right)\left(r^2+a^2\right)\sin^2\vartheta + a^2\frac{\psi}{\Sigma}\sin^4\vartheta\right\} \\
&= -\frac{\Sigma^2}{\Delta}\left\{-\left(r^2+a^2\right)\left(1-\frac{\psi}{\Sigma}\right)\sin^2\vartheta - a^2\frac{\psi}{\Sigma}\sin^4\vartheta\right\} \\
&= -\frac{\Sigma}{\Delta}\left\{-\left(r^2+a^2\right)\Sigma\sin^2\vartheta + \left(r^2+a^2\right)\psi\sin^2\vartheta - a^2\psi\sin^4\vartheta\right\} \\
&= -\frac{\Sigma}{\Delta}\sin^2\vartheta\left\{-\left(r^2+a^2\right)\Sigma + \left(r^2+a^2\right)\psi - a^2\psi\sin^2\vartheta\right\}.
\end{aligned}
\tag{5.105}
$$

Substituting the expression of Σ leads to

$$
\begin{aligned}
-g &= -\frac{\Sigma}{\Delta}\sin^2\vartheta\left\{-\left(r^2+a^2\right)\left[\left(r^2+a^2\right)-a^2\sin^2\vartheta\right]+\left(r^2+a^2-a^2\sin^2\vartheta\right)\psi\right\} \\
&= -\frac{\Sigma}{\Delta}\sin^2\vartheta\left\{-\left(r^2+a^2\right)\left[r^2+a^2\cos^2\vartheta\right]+\left[r^2+a^2\cos^2\vartheta\right]\psi\right\} \\
&= -\frac{\Sigma}{\Delta}\sin^2\vartheta\left(r^2+a^2\cos^2\vartheta\right)\left[-\left(r^2+a^2\right)+\psi\right].
\end{aligned}
\tag{5.106}
$$

The last factor is just $-\Delta$, canceling the delta in the denominator. Finally we obtain

$$
-g = +\Sigma\sin^2\vartheta\left(r^2+a^2\cos^2\vartheta\right),
\tag{5.107}
$$

which does not have anymore a dependence on ψ and Δ, thus it is independent of B.

Exercise 5.8 (Brief Revision of the thin-disk model of Page and Thorne)

We will not give a complete account of the model, which would be out of scope for this chapter. A detailed and very good description is given in [33].

Page and Thorne proposed in [33] a model for an accretion disk. The assumptions are:

(1) The central body determines the external space-time geometry and the influence of the disk is neglected.
(2) The accretion disk lies in the equatorial plane and has infinite extension.
(3) The disk is thin, which means that the thickness $\Delta z = 2h$ of the disk is always much less than the value of the radial distance r.

(4) There exists a time interval Δt, which is small enough that during the external geometry of the central mass practically does not change, but that Δt is large enough that at a given r the total mass that flows inward across r during Δt is large compared with the typical mass contained between r and $2r$.

(5) The ansatz for the energy-momentum tensor is

$$T_{\mu\nu} = \varepsilon_0 \left(1 + \Pi\right) u_\mu u_\nu + t_{\mu\nu} + u_\mu q_\nu + q_\mu u_\nu. \qquad (5.108)$$

where Π is the specific internal energy, $t_{\mu\nu}$ are the tensorial components of the *stress tensor* in averaged rest frame (the stress is generated through interactions between particles in different orbits), u_μ are the components of the 4-velocity and q_μ is the *energy flow* 4-vector orthogonal to the 4-velocity, i.e. $u^\mu q_\mu = 0$. The orthogonality of the energy-flow vector and the velocity means that the particle with a given 4-velocity is moving within the disk, while the energy is emitted perpendicular to the disk.

(6) The particles move in a circular orbital and [33] define the 4-velocity within the disk.

(7) The energy emitted from the disk is only done through the emission of photons, i.e. only radiation exits.

(8) One can neglect the energy and momentum transport from one region of the disk to another one by the photons emitted from the disk's surface.

The ε_0 denotes the mass density in the accretion disk.

The flux function $F(r)$ is defined as (see also Fig. 5.10)

$$F(r) = \langle q^z(r, z = h) \rangle = \langle -q^z(r, z = -h) \rangle. \qquad (5.109)$$

The flux is only emitted into the positive z-direction at the upper surface of the disk, while at the lower surface it is emitted in the negative z-direction.

Also a torque term appears and the time-average torque per unit circumference acting across a cylinder at radius r, due to the stresses within the disk, is

$$W_\varphi^r(r) = \int_{-h}^{+h} \langle t_\varphi^r \rangle dz. \qquad (5.110)$$

The t_φ^r is the φ-r component of the torque tensor defined in (5.108).

In [33] one proceeds in combining the geodesic equation to obtain a relation between the energy and angular momentum

$$E_{|r} = \omega L_{|r}, \qquad (5.111)$$

important for the later solution of the equations coming from conservation laws. The lower index $|r$ denotes the simple derivative with respect to r.

Then, the conservation laws for mass, energy and angular momentum are used to get further equations. For example

$$\nabla (\varepsilon_0 u) = 0, \qquad (5.112)$$

where u_μ are the components of the 4-velocity in the local frame of the particle.

Starting from (5.110), an integral version of the conservation law is obtained by integrating from r to $r + \Delta r$, $z = -h$ until $z = +h$ and in time from t up to $t + \Delta t$. This surface integral results into an equation for $\dot{M}_0$, the change of mass, or the time-averaged rate of the accretion's mass, which is *independent of the radial distance*.

The conservation of energy and angular momentum finally leads to the differential equations

$$(L_z - w)_{|r} = f L_z,$$
$$(E - \omega w)_{|r} = f E, \qquad (5.113)$$

with the definitions

$$f = \frac{-\omega_{|r}}{(E - \omega L_z)^2} \mathscr{I},$$
$$w = \frac{(E - \omega L_z)}{-\omega_{|r}} f = \frac{i}{(E - \omega L_z)} \mathscr{I},$$
$$\mathscr{I} = \int (E - \omega L_z) L_{z|r} dr + \text{const.} \qquad (5.114)$$

The origin of w is the torque transported within the disk, due to the collision of particles in neighboring orbits. In this way, particles in an orbit with a larger rotational frequency transport through collisions energy to particles in an orbit with a lower angular frequency. In standard GR, this implies a transport from lower r to larger r. The f is proportional to the ratio of the flux F with $\dot{M}_0$ and w is proportional to the torque W_φ^r divided by $\dot{M}_0$, i.e. there is a normalization to the unknown value $\dot{M}_0$.

Equation (5.113) can be solved by combining them, giving

$$L_{z|r} - \left(\frac{\omega_{|r} L_z}{(E - \omega L_z)^2} \mathscr{I} + \frac{1}{(E - \omega L_z) \mathscr{I}_{|r}} \right) = f L_z. \qquad (5.115)$$

The interpretation, given in [33], for the different terms in (5.115) is, that the first term describes the angular momentum transported by the mass through the disk, the second term (in the parenthesis) describes the angular momentum transported through the mechanical stresses.

The interesting part is the flux calculated. The normalized flux f is taken, from which F can be deduced. The solution is given in (5.114), where still the integration limits have to be defined. In GR, the lower integration limit is given by the last stable orbit, from which on the energy within the disk is transported further outside, until reaching r, which the position of the emission by light. Thus in standard GR

$$f = -\omega_{|r} \left(E - \omega L_z\right)^{-2} \int_{\text{lso}}^{r} \left(E - \omega L_z\right) L_{z|r} dr, \qquad (5.116)$$

where the index "lso" refers to the *last stable orbit*, also called ISCO (*Innermost Stable Circular Orbit*).

In pc-GR this is changed, because a maximum in ω appears. At the maximum, two neighboring orbitals have nearly the same speed, thus, no energy and torque is transported. From this point on ($r_{\omega_{\max}}$), energy and torque are transported to larger radial distances, because at $r = r_{\omega_{\max}}$ the orbital speed is larger than for $r > r_{\omega_{\max}}$. But also this orbital speed is larger than for $r < r_{\omega_{\max}}$, thus from $r = r_{\omega_{\max}}$ energy and torque is also transported to lower values of r. When the flux in (5.114) is calculated, one has to take this into account, leading to

$$\text{for } r > r_{\omega_{\max}} : \; f = -\omega_{|r} \left(E - \omega L_z\right)^{-2} \int_{r_{\omega_{\max}}}^{r} \left(E - \omega L_z\right) L_{z|r} dr,$$

$$\text{for } r < r_{\omega_{\max}} : \; f = -\omega_{|r} \left(E - \omega L_z\right)^{-2} \int_{r}^{r_{\omega_{\max}}} \left(E - \omega L_z\right) L_{z|r} dr.$$

$$(5.117)$$

This division into two cases is necessary, because otherwise the flux would be negative, due to the derivative in ω in r, which changes sign at the point $r = r_{\omega_{\max}}$.

For more details, please consult [33].

In addition to the assumptions made by [33] we have to include the further assumption that the stresses inside the disk carry angular momentum and energy from faster to slower rotating parts of the disk (see Exercise 5.8). Physically, this can be understood as follows: In standard GR, for two neighboring orbitals, the inner one rotates faster. Colliding with particles just outside, energy is transferred to orbits at larger r, which can be described by torques. In pc GR there s a distinct orbit, namely the one with maximal orbital frequency. There, two neighboring orbitals have nearly the same orbital speed and, thus, no collisions appear and no energy is transferred. Thus,

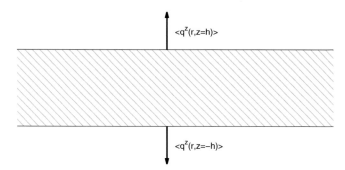

Fig. 5.10 Schematic illustration of an accretion disk. The flux function is indicated by the *arrows* above and below the disk, showing the direction of emission

this radial distance represents a final position from which on, no further energy can be transported. For $r > r_{\omega_{max}}$ energy is transported from this point outwards and for $r < r_{\omega_{max}}$ it is transported inwards. Because at $r = r_{\omega_{max}}$ no collisions take place (or at least are very soft) no heating happens around this point and, thus, a dark ring will appear. In pc-GR, Eq. (5.102) then has to be modified to

$$f = -\omega_{|r}(E - \omega L_z)^{-2} \int_{r_{\omega_{max}}}^{r} (E - \omega L_z)L_{z|r}dr \, ex : thin - disk \, model \quad (5.118)$$

where ω_{max} describes the orbit where the angular frequency ω has its maximum. Equation (5.118) gives a concise way to write down the flux in the two regions (r_{in} describes the inner edge of an accretion disk):

1. $r_{\omega_{max}} < r_{in} \leq r$: This is also the standard GR case, where $\omega_{|r} < 0$ and the flux in Eqs. (5.102) and (5.118) is positive.
2. $r_{in} \leq r < r_{\omega_{max}}$: Here $\omega_{|r} > 0$, but the upper integration limit in (5.118) is smaller than the lower one. Thus there are overall two sign changes and the flux f is positive again.

Thus if we consider a disk whose inner radius is below $r_{\omega_{max}}$, which is the case in the pc-GR model for $a > 0.416 \, m$, Eq. (5.118) guarantees a positive flux function f.

All quantities E, L_z, ω in (5.118) were already computed in the previous section (see also [30]). The angular frequency ω is given in (5.98), E and L_z are given as.

$$L_z^2 = \frac{(g_{03} + \omega g_{33})^2}{-g_{33}\omega^2 - 2g_{03}\omega - g_{00}},$$

$$E^2 = \frac{(g_{00} + \omega g_{03})^2}{-g_{33}\omega^2 - 2g_{03}\omega - g_{00}}. \quad (5.119)$$

Table 5.1 Values for the inner edge of the disks r_{in} in pc-GR for the parameter $B = 64/27\,m^3$

Spin parameter a (m)	r_{in} (m)
0.0	5.24392
0.1	4.82365
0.2	4.35976
0.3	3.81529
0.4	2.99911
0.5 and above	1.334

Unfortunately the derivatives of E and L_z become lengthy in pc-GR and the integral in (5.102) has no analytic solution anymore. Nevertheless it can be solved numerically and thus we are able to modify the original disk model by [33] to include pc-GR correction terms.

5.2.2 Predictions

As shown above, the concept of an ISCO is modified in the pc-GR model. For the following results we used as the inner radius for the disks in the pc-GR case the values depicted in Table 5.1. Values of r_{in} for $a \leq 0.4\,m$ correspond to the modified last stable orbit.

The value of r_{in} for values of a above $0.416\,m$ is chosen slightly above the value $r = (4/3)\,m$. (Remember that in pc-Schwarzschild this is the radial distance where the surface of the star should be, using the minimal value of B). For smaller radii, Eq. (5.98) has no real solutions anymore in the case of $B = (64/27)\,m^3$. The same also holds for general (not necessarily geodesic) circular orbits, where the time component $u^0 = \dfrac{1}{\sqrt{-g_{00} - 2\omega g_{03} - \omega^2 g_{33}}}$ of the particles four-velocity also turns imaginary for radii below $r = (4/3)\,m$ in the case of $B = (64/27)\,m^3$.

We assume that the compact massive object extends up to at least this radius. For all simulations however we did neglect any radiation from the compact object, which is a simplification and has to be addressed in future.

The angular size of the compact object is also modified in the pc-GR case. It is proportional to the radius of the central object [19], which varies in standard GR between 1 and $2\,m$, leading to angular sizes of approximately $10–20\,\mu$as for Sagittarius A*. The size of the central object in pc-GR is fixed at $r = (4/3)\,m$ in the limiting case for $B = (64/27)\,m^3$ thus leading to an angular size of approximately $13\,\mu$as.

5.2.2.1 Images of an Accretion Disk

In Fig. 5.13 we show images of infinite geometrically thin accretion disks according to the model of [33] in certain scenarios. Shown is the bolometric intensity I[erg

$cm^{-2}s^{-1}ster^{-1}$] which is given by $I = \frac{1}{\pi}F$ [26]. To make differences comparable, we adjusted the scales for each value of the spin parameter a to match the scale for the pc-GR scenario. The plots of the Schwarzschild object ($a = 0.0\,m$) and the first Kerr object ($a = 0.3\,m$) use a linear scale whereas the plots for the other Kerr objects ($a = 0.6\,m$ and $a = 0.9\,m$ respectively) use a log scale for the intensity. This is a compromise between comparability between both theories and visibility in each plot. One has to keep in mind, that scales remain constant for a given spin parameter a and change between different values for a.

The overall behavior is similar in GR and pc-GR. The most prominent difference is that the pc-GR images are brighter. An explanation for this effect is the amount of energy which is released for particles moving to smaller radii. This energy is then transported via stresses to regions with lower angular velocity, thus making the disk overall brighter. In Fig. 5.11 we show this energy for particles on stable circular orbits.

It might be at first puzzling, that the fluxes differ significantly for radii above $10\,m$ although here the differences between the pc-GR and standard GR metric become negligible. However the flux f in (5.118) at any given radius r depends on an integral over *all* radii starting from $r_{\omega_{max}}$ up to r. Thus the flux at relatively large radii is dependent on the behavior of the energy at smaller radii, which differs significantly from standard GR.

It is important to stress that the difference in the flux between the standard GR and pc-GR scenarios is also strongly dependent on the inner radius of the disk. This is due to the fact that the values for the energy too are strongly dependent on the radius, see Fig. 5.11a. In Fig. 5.12b we compare the pc-GR and GR case for the same inner radius. There is still a significant difference between both curves but not as strongly as in Fig. 5.12a.

The next significant difference to the standard disk model by [33] is the occurrence of a dark ring in the case of $a \geq 0.416$. This ring appears in the pc-GR case due to the fact that the angular frequency of particles on stable orbits now has a maximum at $r = r_{\omega_{max}} \approx 1.72m$ [30] and the disks extend up to radii below $r_{\omega_{max}}$. At this point the flux function vanishes, which was explained above: Two neighboring orbitals have nearly the same speed and due to this collisions are soft and heating is minimal. As a consequence, a dark ring appears in the accretion disk. Going further inside, the flux increases again, which is a new feature of the pc-GR model. This is the reason of the ring-like structure for $a > 0.416m$. Note that the bright inner ring may be mistaken for second order effects although these do not appear as the disk extends up to the central object.

In Fig. 5.12a we show the radial dependency of the flux function, see (5.102) and (5.118).

For small values of a we still have an ISCO in the pc-GR case and the flux looks similar to the standard GR flux—it is comparable to standard GR with higher values of a. If a increases and we do not have a last stable orbit in the pc-GR case, the flux gets significantly larger and now has a minimum. This minimum can be seen as a dark ring in the accretion disks in Fig. 5.13.

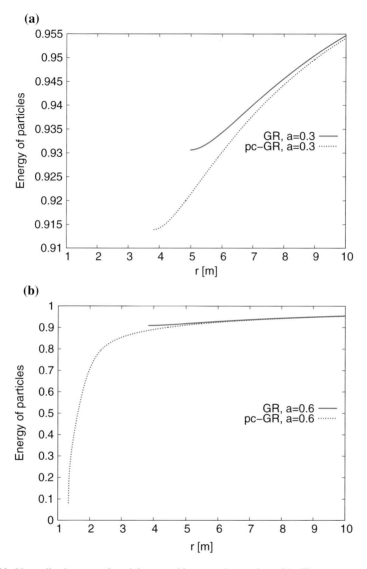

Fig. 5.11 Normalized energy of particles on stable prograde *circular orbits*. The pc-parameter B is set to the critical value of $(64/27)\,m^3$. In the pc-GR case more energy is released as particles move to smaller radii, where the amount of released energy increases significantly in the case where no last stable orbit is present anymore. The lines end at the last stable orbit or at $r = 1.334\,\mathrm{m}$, respectively. **a** $= 0.3\,m$. **b** $= 0.3\,m$

Another feature is the change of shape of the higher order images. For spin values of $a > 0.416m$ the disk extends up to the central object in the pc-GR model, as it is the case for (nearly) extreme spinning objects in standard GR. Therefore no higher order images can be seen in this case. The ringlike shape in Fig. 5.13 is not an image of higher order but still parts of the original disk, as described above.

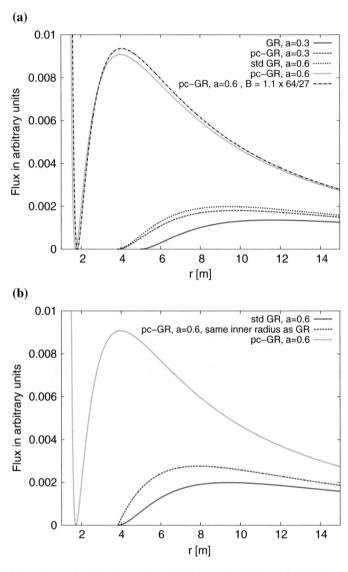

Fig. 5.12 Shown is the flux function f from (5.102) and (5.118) for different values of a (and B). If not stated otherwise $B = (64/27)\,m^3$ is assumed for the pc-GR case. **a** Flux function f for varying spin parameter a and inner edge of the accretion disk. In the standard GR case the ISCO is taken as inner radius, for the pc-GR case see Table 5.1. **b** Dependence of the flux function f on the inner radius of the disk

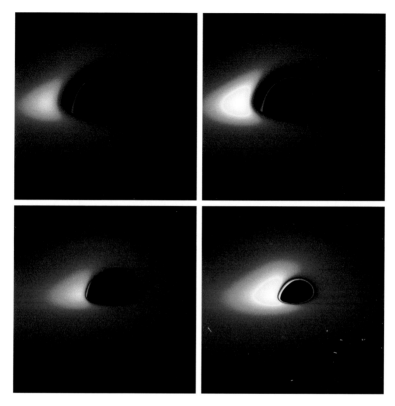

Fig. 5.13 Infinite, counter clockwisely geometrically thin accretion disk around static and rotating compact objects viewed from an inclination of 70°. The *left panel* shows the original disk model by [33]. The *right panel* shows the modified model, including pc-GR correction terms as described in Sect. 5.2.1.1. Scales change between the images. The first row corresponds to the spin parameter $a = 0$, which gives the result for the Schwarzschild case. The second *row* is for $a = 0.9\,m$

5.2.2.2 Emission Line Profiles for the Iron Kα Line

As mentioned earlier, emission line profiles allow to investigate regions of strong gravity. All results in this section share the same parameter values for the outer radius of the disk ($r = 100m$), the inclination angle ($\vartheta = 40°$) and the power law parameter $\alpha = 3$ (as suggested for disks first modeled by [34]). We use this simpler model to simulate emission lines as it is widely used in the literature and thus results are easily comparable. The angle of $\vartheta = 40°$ is just an exemplary value and can be adjusted. As rest energy for the iron Kα line we use 6.4 keV. The inner radius of the disks is determined by the ISCO in GR and by the values in Table 5.1 for pc-GR, and varies with varying values for a. Shown is the flux in arbitrary units. In Figs. 5.14a and 5.16b we compare the influence of the objects spin on the shape of the emission line profile in GR and pc-GR separately. Both in GR and pc-GR we observe the characteristic

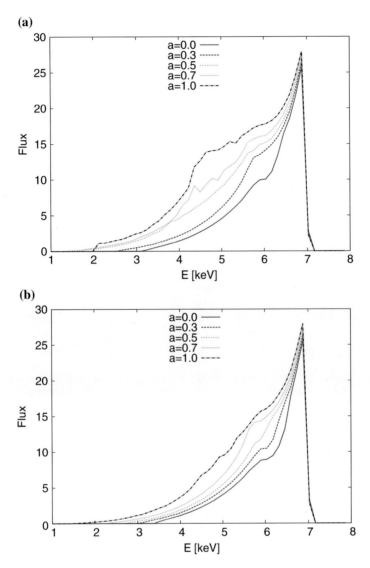

Fig. 5.14 Several line profiles for different values of the spin parameter a. **a** pc-GR. **b** Standard GR

broad and smeared out low energy tail, which grows with growing spin. It is more prominent in the case of pc-GR. The overall behavior is the same in both theories.

A closer comparison of both theories and their differences is given in Figs. 5.15 and 5.16, where we compare the two theories for different values of the spin parameter a. For slow rotating objects (Schwarzschild limit), almost no difference is observable. As the spin grows, we observe an increase of the low energy tail in the pc-GR scenario compared to the GR one. The blue shifted peak however stays nearly the same.

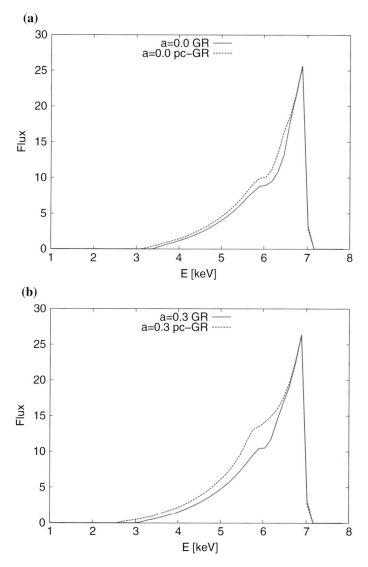

Fig. 5.15 Comparison between theories. The plots are done for parameter values $r = 100\,m$ for the outer radius of the disk, $\vartheta = 40°$ for the inclination angle and $\alpha = 3$ for the power law parameter. The inner radius of the disks is determined by the ISCO and thus varies for varying a. **a** $a = 0.0m$. **b** $a = 0.3m$

If we compare both theories for different values of the spin parameter a they get almost indistinguishable for certain choices of parameters, see Fig. 5.17.

To better understand the emission line profiles we have a look at the redshift in two ways. The redshift can be written as [14]

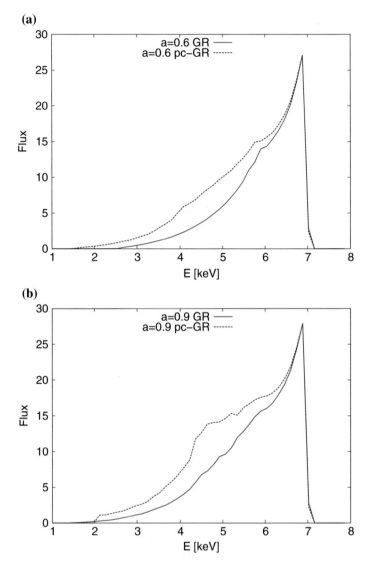

Fig. 5.16 Comparison between theories. The plots are done for parameter values $r = 100\,m$ for the outer radius of the disk, $\vartheta = 40°$ for the inclination angle and $\alpha = 3$ for the power law parameter. The inner radius of the disks is determined by the ISCO and thus varies for varying a. **a** $a = 0.6\,m$. **b** $a = 0.9\,m$

$$g = \frac{1}{u^0_{\mathrm{em}}(1 - \omega\lambda)}, \tag{5.120}$$

where $u^0_{\mathrm{em}} = \frac{1}{\sqrt{-g_{00} - 2\omega g_{03} - \omega^2 g_{33}}}$ is the time component of the emitters four-velocity, ω is the angular frequency of the emitter and λ is the ratio of the emitted photons

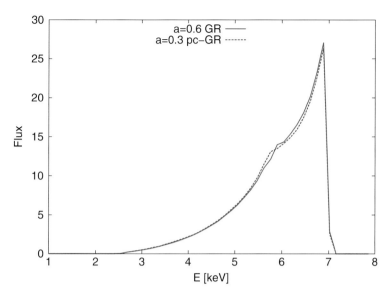

Fig. 5.17 Comparison between theories for different values for the spin parameter. The plot is done for parameter values $r = 100\,m$ for the outer radius of the disk, $\vartheta = 40°$ for the inclination angle and $\alpha = 3$ for the power law parameter. The inner radius of the disks is determined by the ISCO

energy to angular momentum. Reference [35] derived an expression for photons emitted directly in the direction of the emitters movement.

$$\lambda_{\text{cis}} = \frac{-g_{03} - \sqrt{g_{03}^2 - g_{00}g_{33}}}{g_{00}} \tag{5.121}$$

We take this expression and use it to approximate the redshift viewed from an inclination angle ϑ_{obs} as

$$g \approx \frac{1}{u_{\text{em}}^0 (1 - \omega \lambda_{\text{cis}} \sin \vartheta_{\text{obs}})} \tag{5.122}$$

To obtain the full frequency shift one needs in general to know the emission angle of the photon at the point of emission, which can be obtained by using ray-tracing techniques.

5.3 Conclusions

To conclude this chapter, several observables have been calculated, namely

- The orbital frequency of a particle in a circular motion.
- The redshift as a function of the radial distance
- The pictures of accretions disks as a function in the rotational parameter a.

The main findings are

- A orbital frequency is lower than in GR, thought the differences become apparent only near the Schwarzschild radius.
- There is a maximal orbital frequency in pc-GR which decreases from this point on toward smaller r.
- The redshift in the equatorial plane is similar to GR, but the function in r is shifted to smaller r and the redshift is not infinity but becomes very large. On the poles this redshift is smaller.
- In the accretion disk a dark ring is predicted to appear, due to the presence of the maximal orbital frequency.

It seems, that galactic "black holes" are not in agreement to observation, because QPO's do show an orbital frequency suggesting large radial distances, while the redshift of Fe-K lines suggest small radial distances. In pc-GR there is a good agreement.

References

1. ALMA observatory, http://www.almaobservatory.org/es
2. S. Gillessen, R. Genzel, T.K. Fritz, E. Quataert, C. Alig, A. Burkert J. Cuadra, F. Eisenhauer, O. Pfuhl, K. Dodds-Eden, C.F. Gammie, T. Ott, A gas cloud on its way towards the supermassive black hole at the Galactic Centre. Nature **481**(51) (2012)
3. B. Aschenbach, N. Grosso, D. Porquet, P. Predehl, X-ray flares reveal mass and angular momentum of the Galactic Center black hole. A&A **417**, 71 (2004)
4. R. Genzel, A. Eckart, R. Schödel, T. Ott, T. Alexander, F. Lacombe, D. Rouan, B. Aschenbach, Near-infrared flares from accreting gas around the supermassive black hole at the Galactic Centre. Nature **425**, 934 (2003)
5. K. Iwasawa, G. Miniutti, A.C. Fabian, Flux and energy modulation of redshifted iron emission in NGC 3516: implications for the black hole mass. MNRAS **355**, 1073 (2004)
6. T.M. Belloni, A. Sanna, M. Mendz, High-frequency quasi periodic oscillations in black hole binaries. MNRAS **426**, 1701 (2012)
7. R.I. Hynes, D. Steeghs, J. Casares, P.A. Charles, K.O.'Bri2n, The distance and interstellar sight line to GX 339-4. ApJ **609**, 317 (2004)
8. J.M. Miller, J.M. Reynolds, A.C. Fabian et al., Initial measurements of black hole spin in GX 339–4 from Sazuka spectroscopy. ApJL **679**, L113 (2008)
9. R.C. Reis, A.C. Fabian, R.R. Ross, G. Miniutti, J.M. Miller, C. Reynolds, A schematic look at the very high and low/hard state of GX 339–4: Constraining the black hole spin with a new reflection model. MNRAS **387**, 1489 (2008)
10. J. Steiner, J. McClintock, G. Jeffrey et al., in *Measuring the Spin of Accreting Black Holes*, 38th COSPAR Scientific Assembly, Bremen, Germany, 18–15 July 2010
11. D. Lai, W. Fu, D. Tsang, J. Horak, C. Ya, High-frequency QPO's and overstable oscillations of black-hole accretion disks. Proceedings of the International Astronomical Union **8S**(290), 57 (2012)
12. C.M. Will, The confrontation between General Relativity and experiment. Living Rev. Relativ. **9**, 3 (2006)
13. R.A. Hulse, J.H. Taylor, Discovery of a pulsar in a binary system. Astrophys. J. **195**, L51 (1975)
14. C. Fanton, M. Calvaniand F. de Felice, A. Cadez, in *Detecting Accretion Disks in Active Galactic Nuclei. Publications of the Astronomical Society of Japan, 1997*, p. 159

15. A. Müller, M. Camenzind, Relativistic emission lines from accreting black holes. A&A **413**, 861 (2004)
16. R. Adler, M. Bazin, M. Schiffer, *Introduction to General Relativity*, 2nd edn. (McGraw Hill, New York, 1975)
17. G. Caspar, T. Schönenbach, P.O. Hess, M. Schäfer, W. Greiner, Pseudo-Complex General Relativity: Schwarzschild, Reissner-Nordstrøm and Kerr Solutions. Int. J. Mod. Phys E. **21**, 1250015 (2012)
18. C.W. Misner, K.S. Thorne, J.A. Wheeler, *Gravitation* (Freeman & Co., San Francisco, 1973)
19. A. Müller, Experimental evidence of black holes. School on Particle Physics, Gravity and Cosmology, 21.08–02.09.2006 in Dubrovnik. in *Proceedings of Science 017* (2007)
20. D.C. Wilkins, Bound geodesics in the Kerr metric. Phys. Rev. D **5**, 814 (1972)
21. B. Anderson, J. Jackson, M. Sitharam, Descartes' rule of signs revisited. Amer. Math. Monthly **105**, 447 (1998)
22. Mathematica, Version 9.0. Wolfram Research, Inc. Champaign, Illinois, 2012
23. J.M. Bardeen, W.H. Press, S.A. Teukolsky, Rotating Black holes: locally nonrotating frames, energy extraction and scalar synchrotron radiation. ApJ **178**, 347 (1972)
24. T. Boller, A. Müller, Observational tests of the pseudo-complex theory of GR using black hole candidates, in textitNuclear Physics: Present and Future, FIAS Lecture Series ed. by W. Greiner (2014), p. 245
25. C. Bambi, D. Malafarina, K-α iron line profile from accretion disks around regular and singular exotic compact objects. Phys. Rev. D **88**, 064022 (2013)
26. F.H. Vincent, T. Paumard, E. Gourgoulhon, G. Perrin, GYOTO: a new general relativistic ray-tracing code. Class. Quantum Gravity **28**(22), 225011 (2011)
27. T. Schoenenbach, Tests erweiteter Kerr und Schwarzschild-Metriken im Rahmen der Beschreibung von Teilchenorbits, Akkretionscheiben und Gravitationswellen. Ph.D. thesis, J.W. von Goethe University, Frankfurt am Main, Germany, 2015
28. J. Levin, G. Perez-Giz, A periodic table for black hole orbits. Phys. Rev. D **77**, 103005 (2008)
29. B. Carter, Global structure of the Kerr Family of gravitational fields. Phys. Rev. **174**, 1559 (1968)
30. T. Schönenbach, G. Caspar, P.O. Hess, T. Boller, A. Müller, M. Schäfer, Walter Greiner, Experimental tests of pseudo-complex general relativity. MNRAS **430**, 2999 (2013)
31. T. Schönenbach, G. Caspar, P.O. Hess, T. Boller, A. Müller, M. Schäfer, W. Greiner, Ray-tracing in pseudo-complex General Relativity. MNRAS **442**, 121 (2014)
32. G.B. Rybicki, A.P. Lightman, *Radiative Processes in Astrophysics* (Wiley-VCH Verlag GmbH & Co, KGaA, 2004)
33. D.N. Page, K.S. Thorne, Disk-accretion onto a black hole. Time-averaged structure of accretion disk. Astrophys. J. **191**, 499 (1974)
34. N.I. Shakura, R.A. Sunyaev, Black holes in binary systems. Observational appearance. A&A **24**, 337 (1973)
35. S. Cisneros, G. Goedecke, C. Beetle, M. Engelhardt, On the Doppler effect for light from orbiting sources in Kerr-type metrics, 2012, arXiv:gr-qcl1203.2502

Chapter 6
Neutron Stars Within the Pseudo-complex General Relativity

For convenience, in this chapter we use natural units $c = \kappa = 1$, which are commonly used in the field of neutron stars.

Until here, the pc-metric *outside* a mass distribution was considered. However, one of the most interesting effects is encountered when a mass distribution is added, for example *inside* a star. As we have seen up to now, sensible differences between pc-GR and standard GR do only appear near the Schwarzschild radius. Therefore, considering normal stars will not show any differences. The situation is different when compact objects like neutron stars are considered. Neutron stars are one possible outcome of the collapse of a massive star. Once the nuclear fuel is consumed, massive stars undergo a supernova explosion where the outer layers of the star are blown off. Gravity makes the inner region to collapse in such a way that protons and electrons combine to form neutrons.

In some cases sufficient matter remains in the central object, which is formed during the collapse, that it continues to collapse to a black hole, at least according to the standard theory. The existence of black holes has become commonly accepted despite the fact that the existence of event horizons cannot be proved from observational data [1].

In the former chapters we stressed our philosophical point of view, that *in a theory no singularities are allowed to appear*, not even nearby coordinate singularities. If such singularities are present, it is rather a sign of the incompleteness of such a theory. We argue that no black hole should exist and that large masses, which in the standard theory would form a black hole, will resemble huge, heavily compressed stars. If these stars have an inner structure as a neutron star has to be determined yet. We can not exclude exotic stars, like quark stars or stars with an internal matter structure not known up to now! This is beyond the current theory of GR and pc-GR.

We have proposed an alternative to the black hole, namely that the Einstein equations have an additional, repulsive contribution due to the accumulation of dark energy. Dark energy is used in order to explain the present phase of acceleration of

© Springer International Publishing Switzerland 2016   183
P.O. Hess et al., *Pseudo-Complex General Relativity*,
FIAS Interdisciplinary Science Series, DOI 10.1007/978-3-319-25061-8_6

the universe and there is no argument to exclude the accumulation of dark energy around masses. We gave arguments in favor of it in the former chapters and identified the accumulation of the dark energy due to vacuum fluctuations. It is reasonable to expect that this component, commonly called dark energy, also affects smaller scale phenomena like the gravitational collapse of bound objects.

Let us first shortly mention some contributions, which also include dark energy, in order to emphasize later on the differences to our theory: A model of dark energy stars has been formulated in [2] where the event horizon is proposed to be a quantum phase transition analogous to the critical point of a Bose fluid. Another model with similar characteristics is the gravitational vacuum star ("gravastar") model [3]. A phase transition for the quantum vacuum takes place near the location where the horizon is expected to be. The model is a static, spherical symmetric five layer solution of the GR equations. It assumes the existence of a compact object with an interior de Sitter geometry and an equation of state $p = -\varepsilon$ which is matched to a finite thickness shell with $p = \varepsilon$. The latter is then matched to an exterior Schwarzschild vacuum solution and the three regions are connected by two thin anisotropic layers with distributions of surface tension and surface energy density. The solution presents no singularities and no event horizons. A simplified version of this model is studied in [4] where the thick shell and the two thin ones are combined into a single thin one which matches a de-Sitter interior with a vacuum Schwarzschild exterior. The simplicity of this model allows for a full dynamical analysis where stability is found for some physically reasonable equations of the state of the thin shell. In [5] similar models exhibiting a continuous pressure profile, without the presence of thin shells are studied. It is found that gravastars cannot be perfect fluids, the presence of anisotropic pressures is unavoidable. Related models have also been analyzed in [6, 7] and a generalization of the gravastar picture has been given in [8].

As pointed out in [9] the conservation of energy implies a constant energy density for a fluid of the cosmological constant type in the absence of matter or other fields. Thus, dark energy must be coupled to matter in order to form a condensate. This can be achieved by a direct proportionality between these two components. In our theory such a relation has been chosen but negative values of the coupling parameter have been taken. The theoretical possibility of the existence of negative energy densities have been described in [10–13] where vacuum fluctuations have been discussed in presence of matter using different approaches for the vacuum. A semi-classical study of the gravitational collapse outcome has been carried in [14] where the standard black hole picture is questioned as the appropriate end point of a realistic collapse. The possibility of negative energy is also mentioned.

In the present chapter we mainly copy the results of our calculations for neutron stars, which were published in [15]. For simplicity, non-rotating stars are considered, i.e., the pc-Schwarzschild solution with mass present (please, refer to Chap. 3).

6.1 Theoretical Background

Within the pc-GR theory, the Einstein equations include an extra term associated to the nature of space-time itself (see Chap. 3). This term is believed to halt the collapse of matter distributions avoiding the standard GR predictions of black hole creation as the final stage [14]. In order to proceed within a familiar framework, we represent this term by an energy momentum tensor $(T^\Lambda)_{\mu\nu}$ which describes its contribution. The physical origin of this term may arise from micro-scale phenomena where vacuum fluctuations could become considerable under certain conditions [14].

6.1.1 Interior Region

With the presence of a standard matter distribution characterized by an energy momentum tensor $(T^m)_{\mu\nu}$ the real projection (which has to be done within the framework of pc-GR) of the modified Einstein equations takes the form

$$\mathscr{R}_\nu{}^{\mu i} - \frac{1}{2} g_\nu{}^{\mu i} \mathscr{R}^i = 8\pi \left(T^\Lambda\right)_\nu{}^{\mu i} + 8\pi \left(T^m\right)_\nu{}^{\mu}, \tag{6.1}$$

(i stands for interior) where $\mathscr{R}_\nu{}^\mu$ denotes the real projection of the Ricci tensor, $\mathscr{R}$ the real projection of the Ricci scalar and $g_\nu{}^\mu$ the corresponding projection of the metric. Within the scope of this chapter, only static spherical symmetric objects will be analyzed:

$$ds^2 = -e^{\nu_i(r)} c^2 dt^2 + e^{\lambda_i(r)} dr^2 + r^2 (d\vartheta^2 + \sin^2 \vartheta d\varphi^2). \tag{6.2}$$

For both components, the isotropic perfect fluid assumption will be considered as well: (see Exercise 6.1),

$$\left(T^\Lambda\right)_\nu{}^{\mu i} = \begin{bmatrix} -\varepsilon_{\Lambda i} & 0 & 0 & 0 \\ 0 & p_{\Lambda i} & 0 & 0 \\ 0 & 0 & p_{\Lambda i} & 0 \\ 0 & 0 & 0 & p_{\Lambda i} \end{bmatrix}, \tag{6.3}$$

$$\left(T^m\right)_\nu{}^{\mu} = \begin{bmatrix} -\varepsilon_m & 0 & 0 & 0 \\ 0 & p_m & 0 & 0 \\ 0 & 0 & p_m & 0 \\ 0 & 0 & 0 & p_m \end{bmatrix}, \tag{6.4}$$

where ε and p denote energy density and pressure respectively. Given the symmetry, both are going to be functions only of the radius. The Ricci tensor and the Ricci scalar are expressed as usual [16] (see also Chap. 3 on the pc-Schwarzschild metric):

$$\mathcal{R}_{00} = \frac{1}{2} e^{\nu - \lambda} \left[\nu'' + \frac{\nu'^2}{2} - \frac{\nu' \lambda'}{2} + \frac{2\nu'}{r} \right], \tag{6.5}$$

$$\mathcal{R}_{11} = -\frac{1}{2} \left[\nu'' + \frac{\nu'^2}{2} - \frac{\nu' \lambda'}{2} - \frac{2\lambda'}{r} \right], \tag{6.6}$$

$$\mathcal{R}_{22} = -\left(e^{-\lambda} r \right)' + 1 - r \left(\frac{\lambda' + \nu'}{2} \right) e^{-\lambda}, \tag{6.7}$$

$$\mathcal{R}_{33} = \sin^2 \vartheta \, \mathcal{R}_{\vartheta\vartheta} \tag{6.8}$$

$$\mathcal{R} = -e^{-\lambda} \left[\nu'' - \frac{\nu' \lambda'}{2} + \frac{\nu'^2}{2} + \frac{2\nu'}{r} - \frac{2\lambda'}{r} + \frac{2}{r^2} \right] + \frac{2}{r^2}. \tag{6.9}$$

("i", for "interior", will be removed temporarily and then added at final expressions). Using expressions (6.3) and (6.4) together with the previous expressions for the Ricci tensor and scalar, Eq. (6.1) can be split into components. After some minor algebraical work (see Exercise 6.2) the following expression is obtained for the temporal one:

$$e^{-\lambda} \left[-\frac{\lambda'}{r} + \frac{1}{r^2} \right] - \frac{1}{r^2} = -8\pi \varepsilon_m - 8\pi \varepsilon_\Lambda. \tag{6.10}$$

Proceeding analogously a similar relation is obtained for the radial component:

$$e^{-\lambda} \left[\frac{\nu'}{r} + \frac{1}{r^2} \right] - \frac{1}{r^2} = \pi p_m + 8\pi p_\Lambda, \tag{6.11}$$

and also for the angular one:

$$e^{-\lambda} \left[\frac{\nu''}{2} - \frac{\nu' \lambda'}{4} + \frac{\nu'^2}{4} + \frac{\nu'}{2r} - \frac{\lambda'}{2r} \right] = 8\pi \kappa p_m + 8\pi p_\Lambda. \tag{6.12}$$

After some transformations, Eq. (6.10) can be used to express the radial metric coefficient:

$$e^{-\lambda_i(r)} = 1 - \frac{2m_m(r)}{r} + \frac{2m_{\Lambda i}(r)}{r}, \tag{6.13}$$

where m_m and $m_{\Lambda i}$ are defined by:

$$m_m(r) = 4\pi \int_0^r r'^2 \varepsilon_m(r') \, dr', \tag{6.14}$$

$$m_{\Lambda i}(r) = -4\pi \int_0^r r'^2 \varepsilon_{\Lambda i}(r') \, dr'. \tag{6.15}$$

Exercise 6.1 (Form of the energy-momentum tensor of a perfect isotropic fluid)

Problem. Show that for an isotropic fluid the energy-momentum tensor has the structure as given in (6.3) and (6.4). Use a system were the only non-vanishing component of the 4-velocity is u^0.

Solution.
The dispersion relation is given by

$$g_{\mu\nu}u^\mu u^\nu = -1. \tag{6.16}$$

In a system were only u^0 is different from zero, we have

$$g_{00}\left(u^0\right)^2 = -1, \tag{6.17}$$

from which follows that, considering that g_{00} is negative,

$$u^0 = \frac{1}{\sqrt{|g_{00}|}}. \tag{6.18}$$

For the u_0 follows

$$u_0 = g_{00}u^0 = -\sqrt{|g_{00}|}. \tag{6.19}$$

The energy-momentum tensor for an isotropic ideal fluid is given by

$$T_\mu^\nu = (\varepsilon + p)\, u_\mu u^\nu + pg_\mu^\nu. \tag{6.20}$$

with $g_\mu^\nu = \delta_\mu^\nu$.
With this, for T_0^0 and T_1^1 we have

$$T_0^0 = (\varepsilon + p)\, u_0 u^0 + p = (\varepsilon + p)\,(-1) + p = -\varepsilon,$$
$$T_1^1 = (\varepsilon + p)\, u_1 u^1 + p = p, \tag{6.21}$$

where for T_1^1 we used that $u_1 = u^1 = 0$. For the remaining two components the same holds

Exercise 6.2 (Some proofs)

Problem. Proof (6.10)–(6.13).

Solution. The left hand side of the Einstein equations, with $\mu = \nu$ is given for the temporal part by

$$
\begin{aligned}
G^0_0 &= \mathscr{R}^0_0 - \frac{1}{2}\mathscr{R} = \frac{1}{2}\left(\mathscr{R}^0_0 - \mathscr{R}^1_1 - \mathscr{R}^2_2 - \mathscr{R}^3_3\right) \\
&= \frac{1}{2}\left(g^{00}\mathscr{R}_{00} - g^{11}\mathscr{R}_{11} - g^{22}\mathscr{R}_{22} - g^{33}\mathscr{R}_{33}\right) \\
&= -\frac{1}{2}\left(e^{-\nu}\mathscr{R}_{00} + e^{-\lambda}\mathscr{R}_{11} + \frac{1}{r^2}\mathscr{R}_{22} + \frac{1}{r^2\sin^2\vartheta}\mathscr{R}_{33}\right).
\end{aligned}
\tag{6.22}
$$

The last two terms can be joint because $\mathscr{R}_{33} = \sin^2\vartheta\,\mathscr{R}_{22}$.

The same for $\mu = \nu = 1, 2$:

$$
\begin{aligned}
G^1_1 &= \frac{1}{2}\left(e^{-\lambda}\mathscr{R}_{11} + e^{-\nu}\mathscr{R}_{00} - \frac{1}{r^2}\mathscr{R}_{22} - \frac{1}{r^2\sin^2\vartheta}\mathscr{R}_{33}\right), \\
G^2_2 &= \frac{1}{2}\left(\frac{1}{r^2}\mathscr{R}_{22} + e^{-\nu}\mathscr{R}_{00} - e^{-\lambda}\mathscr{R}_{11} - \frac{1}{r^2\sin^2\vartheta}\mathscr{R}_{33}\right).
\end{aligned}
\tag{6.23}
$$

Let us start with the 00-component. Substituting into it (6.5)–(6.8), we obtain

$$
\begin{aligned}
-\frac{1}{2}\Bigg\{ & e^{-\lambda}\frac{1}{2}\left(\nu'' + \frac{(\nu')^2}{2} - \frac{\nu'\lambda'}{2} + \frac{2\nu'}{r}\right) \\
& -e^{-\lambda}\frac{1}{2}\left(\nu'' + \frac{(\nu')^2}{2} - \frac{\nu'\lambda'}{2} - \frac{2\lambda'}{r}\right) \\
& -\frac{2}{r^2}\left(re^{-\lambda}\right)' + \frac{2}{r^2} - \frac{2}{r}\left(\frac{\lambda'+\nu'}{2}\right)e^{-\lambda}\Bigg\},
\end{aligned}
\tag{6.24}
$$

were we made use of the relation between R_{33} and R_{22}. Most terms cancel and what remains is

$$
\frac{1}{r^2}\left(re^{-\lambda}\right)' - \frac{1}{r^2} = e^{-\lambda}\left[-\frac{\lambda'}{r} + \frac{1}{r^2}\right] - \frac{1}{r^2}.
\tag{6.25}
$$

This gives the left hand side of (6.10). The right hand side is simply obtained, using the definition of the energy momentum tensors of mass and dark energy.

We proceed in a similar way for the radial component (6.11): The component $G_1{}^1$ of the Einstein tensor is given by

$$G_1{}^1 = \frac{1}{2}\left(\mathscr{R}_1{}^1 - \mathscr{R}_0{}^0 - \mathscr{R}_2{}^2 - \mathscr{R}_3{}^3\right)$$

$$= \frac{1}{2}\left(e^{-\lambda}\mathscr{R}_{11} + e^{-\nu}\mathscr{R}_{00} + \frac{1}{r^2}\mathscr{R}_{22} + \frac{1}{r^2\sin^2\vartheta}\mathscr{R}_{33}\right). \qquad (6.26)$$

Using $\mathscr{R}_{33} = \sin^2\vartheta\, R_{22}$ and substituting into it (6.5)–(6.8), we get

$$-\frac{e^{-\lambda}}{4}\left(\nu'' + \frac{(\nu')^2}{2} - \frac{\nu'\lambda'}{2} - \frac{2\lambda'}{r}\right)$$

$$+\frac{e^{-\lambda}}{4}\left(\nu'' + \frac{(\nu')^2}{2} - \frac{\nu'\lambda'}{2} + \frac{2\nu'}{r}\right)$$

$$+\frac{2}{2r^2}\left[\left(e^{-\lambda}r\right)' - 1 + r\left(\frac{\lambda'+\nu'}{2}\right)e^{-\lambda}\right], \qquad (6.27)$$

where in the last term the factor 2 in the numerator originates in the symmetry between the 2 and 3 diagonal component. Several terms cancel in (6.27) and what remains is

$$\frac{e^{-\lambda}}{r}\left(\lambda' + \nu'\right) + e^{-\lambda}\left(-\frac{\lambda'}{r} + \frac{1}{r^2}\right) - \frac{1}{r^2} = e^{-\lambda}\left(\frac{\nu'}{r} + \frac{1}{r^2}\right) - \frac{1}{r^2},$$

$$(6.28)$$

which is the left hand side of (6.11). The right hand side is a direct result of the definition of the energy-momentum tensors for the baryonic and the dark energy mass.

What remains is (6.12): The Einstein tensor element $G_2{}^2$ is given by

$$G_2{}^2 = \frac{1}{2}\left(\mathscr{R}_2{}^2 - \mathscr{R}_0{}^0 - \mathscr{R}_1{}^1 - \mathscr{R}_3{}^3\right)$$

$$= \frac{1}{2}\left(\frac{1}{r^2}\mathscr{R}_{22} + e^{-\nu}\mathscr{R}_{00} - e^{-\lambda}\mathscr{R}_{11} - \frac{1}{r^2\sin^2\vartheta}\mathscr{R}_{33}\right). \qquad (6.29)$$

Using $\mathscr{R}_{33} = \sin^2\vartheta\,\mathscr{R}_{22}$, we get

$$G_2{}^2 = \frac{1}{2} - \left(e^{-\lambda}\mathscr{R}_{11} + e^{-\nu}\mathscr{R}_{00}\right). \qquad (6.30)$$

Substituting into it (6.5) snf (6.6), we obtain

$$\frac{e^{-\lambda}}{4}\left(\nu'' + \frac{(\nu')^2}{2} - \frac{\nu'\lambda'}{2} - \frac{2\lambda'}{r}\right) + \frac{e^{-\lambda}}{4}\left(\nu'' + \frac{(\nu')^2}{2} - \frac{\nu'\lambda'}{2} + \frac{2\nu'}{r}\right)$$

$$= \frac{e^{-\lambda}}{2}\left[\nu'' + \frac{(\nu')^2}{2} - \frac{\nu'\lambda'}{2} + \frac{(\nu'-\lambda')}{r}\right], \qquad (6.31)$$

which is the left hand side of (6.12). The right hand side is again the direct result of the definitions of the energy momentum tensors for the baryonic and dark energy part.

In order to obtain (6.13), we start from (6.10), multiply by r^2, use $e^{-\lambda}$ $(-r\lambda' + 1) = (re^{-\lambda})'$ and integrate from 0 to r, which gives

$$\left(re^{-\lambda}\right) - r = -\frac{8\pi}{c^2}\int_0^r r^2\left(\varepsilon_m + \varepsilon_{\Lambda i}\right)dr = -2m_m + 2m_{\Lambda i}, \quad (6.32)$$

where the definition of (6.14) and (6.15) have been used.

Dividing this equation by r gives the desired result of (6.13).

The temporal metric coefficient can be related to the radial one if we subtract (6.11) from (6.10) and make subsequent changes. The obtained relation follows:

$$e^{\nu_i(r)} = e^{-\lambda_i(r)}e^{\frac{f_i(r)+C_i}{2}}, \qquad (6.33)$$

where $f_i(r)$ is defined by:

$$f_i(r) = 8\pi\int_0^r r'e^{\lambda_i(r')}$$
$$\times\left(\varepsilon_m(r') + \varepsilon_{\Lambda i}(r') + p_m(r') + p_{\Lambda i}(r')\right)dr', \qquad (6.34)$$

and the constant C_i is used for the sake of continuity (see Sect. 6.1.5). Hydrostatic equilibrium equations are derived by going further with the algebraical transformations and including (6.12). They can be expressed as the following coupled system of differential equations:

$$\frac{dp_m}{dr} = -\frac{(\varepsilon_m(r) + p_m(r))}{r[r - 2m_m(r) + 2m_\Lambda(r)]}$$
$$\times\left[m_m(r) - m_\Lambda(r) + 4\pi r^3(p_\Lambda(r) + p_m(r))\right], \qquad (6.35)$$

$$\frac{dp_{\Lambda i}}{dr} = -\frac{\left(\varepsilon_{\Lambda i}(r) + p_{\Lambda i}(r)\right)}{r\left[r - 2m_m(r) + 2m_{\Lambda i}(r)\right]}$$
$$\times \left[m_m(r) - m_{\Lambda i}(r) + 4\pi r^3 (p_{\Lambda i}(r) + p_m(r))\right]. \quad (6.36)$$

The last equations were derived in Chap. 3 of this book, leading to (3.53) of this chapter. In order to obtain the above equations, one has to double (3.53), one for the baryonic and the other for the dark energy pressure and on the right hand side of (51) the pressure by the sum of the baryonic and the dark energy pressure. One also assumes that the only interaction between the two fluids is the gravitation, allowing this separation.

When one goes back to standard GR, Eqs. (6.35) and (6.36) reduce to the standard Tolman-Oppenheimer-Volkoff (TOV) equations [17, 18]. These equations, together with the derivative of (6.14), (6.15) form a system which can be closed only if another couple of relations are given. The current study considers an equation of state for the standard matter (i.e. a relation between p_m and ε_m) and a linear coupling between baryonic and Λ-term energy densities.

6.1.2 Equation of State for Standard Matter

In recent decades many different equations of state for star matter have been developed and employed in the study of compact star properties. These equations of state originate from different model assumptions for ground state matter and dense matter. Thus, in case the three-flavor quark matter is the real ground state of strongly interacting matter, the correct equation of state would be that of a strange quark star. The more conventional picture assumes nuclear matter to constitute the ground state of strongly interacting matter. In this scenario compact stars consist of hadrons and, depending on parameters, might contain a mixed phase of quarks and hadrons in the core of the star, which is then termed hybrid star. The most simple descriptions of neutron stars only take into account neutrons and some protons and electrons as degrees of freedom. However, as is known from hyper nuclear physics at least the Λ baryon is bound with about 30 MeV in nuclear matter, thus exhibiting attraction in matter, and most likely the Ξ hyperon is bound as well. Therefore any realistic hadronic model of neutron star matter should also contain hyperons as degrees of freedom that exhibit realistic optical potentials. An approach that fulfills those requirements was discussed for the case of star matter in [19]. The underlying framework is a flavor-SU(3) effective chiral model that includes all low-lying baryonic and mesonic SU(3) multiplets including the strangeness degree of freedom, thus allowing for the possibility of hyperons in the star. Details of the model are described in various publications [19–22]. Baryon masses are generated by their coupling to the scalar mesonic fields. As the fields change in the dense medium so do the baryon masses, effectively generating a scalar attraction. The effective masses m_i^* read

$$m_i^* = g_{i\sigma}\sigma + g_{i\zeta}\zeta + g_{i\delta}\delta + \delta m_i, \tag{6.37}$$

including couplings to the scalar fields (σ, δ, ζ) whose expectation values represent the scalar quark condensates, i.e. $\sigma \sim \langle \bar{u}u + \bar{d}d \rangle$, $\zeta \sim \langle \bar{s}s \rangle$, and $\delta \sim \langle \bar{u}u - \bar{d}d \rangle$, with an additional mass term δm_i that breaks chiral and SU(3) symmetry explicitly. The various couplings contained in the equation result from the SU(3) coupling scheme [20]. The scalar fields σ and ζ attain non-zero vacuum expectation values due to their self interaction [19]. Following (6.37) this generates the baryonic masses in the vacuum, while the change of the scalar fields at finite density or temperature reduces the masses and thus generate scalar attraction. The corresponding effective baryonic chemical potentials read

$$\mu_i^* = \mu_i - g_{i\omega}\omega - g_{i\rho}\rho - g_{i\varphi}\varphi, \tag{6.38}$$

where the different fields ω, ρ and φ are the analogous vector fields to the scalar σ, δ, and ζ, respectively. These fields (in mean-field approximation) can have non-zero values in dense matter, which shifts the effective chemical potentials of the particles, generating a repulsive interaction. As in other non-chiral relativistic nuclear models the interplay between the scalar attraction and vector repulsion leads to the binding of nuclear matter, yielding realistic values for the properties of saturated matter [19]. The maximum star mass in this approach for standard general relativity is $M = 2.06 M_\odot$ [19]. Thus, the model results are in very good agreement with the recent observations of neutron stars with 2 solar masses [23, 24].

The model equation of state (EoS), i.e. the relation between pressure p and energy density ε, that is the defining quantity for the star masses and radii can be seen in Fig. 6.1. For comparison the stiffest EoS with $p = \varepsilon$ and an EoS for free masses particles $p = \varepsilon/3$ is shown. The latter one would correspond to a bag model equation of state for quark stars, which in addition would also include a constant bag pressure that can be fixed to shift the EoS and thus change the resulting star masses and radii.

6.1.3 Equation of State for the Λ-term

As mentioned before, a quantum or semi-classical theory is needed in order to properly describe the vacuum fluctuations in the presence of matter, see [10–13] where different vacuum approaches are used on a Schwarzschild background. Within a classical theory one can only but assume an expression for the energy density. For the sake of simplicity a linear relation between ε_Λ and ε_m is proposed:

$$\varepsilon_\Lambda = \alpha \varepsilon_m. \tag{6.39}$$

A relation of this kind is employed in [9] where positive values for α are used and the linear relation does not hold for the whole distribution but only for some range where the energy density is higher than some critical value. Here five negative values

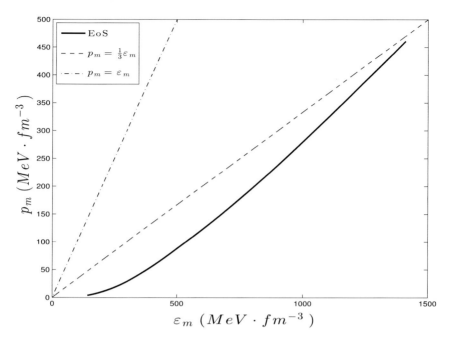

Fig. 6.1 Equation of State for star matter resulting for the standard matter component. Also shown are the EoS for $p = \varepsilon$ and $p = \frac{1}{3}\varepsilon$

of α are studied and the relation will be used for all values of ε_m. These values also satisfy $|\alpha| < 1$ in order to maintain a positive total energy density.

It is expected that the coupling between matter and dark energy, as given in (6.39) is too simplistic. However, this is all what we can do for the moment, due to the absence of a quantized theory of gravitation. A future possibility is to apply semi-classical Quantum Mechanics, i.e. with a fixed back-ground metric, and resolve the Einstein equations in the presence of mass. This is a difficult endeavor and would break the scope of this book, also because these calculations have not been done yet.

6.1.4 Exterior Shell

The Λ-term pressure p_Λ does not vanish at $r = R_0$, it decreases from that point until it does vanish at some larger distance. As previously mentioned, pc-GR predicts the presence of a term (expressed as an energy momentum tensor) even when normal baryonic matter is absent (e stands for exterior):

$$\mathscr{R}_\nu^{\mu e} - \frac{1}{2}g_\nu^{\mu e}\mathscr{R}^e = 8\pi T^{\Lambda\,\mu e}_{\quad\nu}. \tag{6.40}$$

The failure of isotropic pressures within the gravastar context has been discussed in [5] where a continuous radial pressure profile has also been employed avoiding the need of introducing infinitesimally thin shells [9, 25–27]. Following these ideas our exterior shell is chosen to have a non-isotropic energy momentum tensor:

$$
T^{\Lambda\,\mu e}{}_{\nu} =
\begin{bmatrix}
-\varepsilon_{\Lambda e} & 0 & 0 & 0 \\
0 & p_{\Lambda re} & 0 & 0 \\
0 & 0 & p_{\Lambda\vartheta e} & 0 \\
0 & 0 & 0 & p_{\Lambda\vartheta e},
\end{bmatrix}
\tag{6.41}
$$

where r and ϑ stand for radial and tangential respectively. If the analysis given between (6.1) and (6.36) is reproduced for the former energy momentum tensor, analogous expressions to the interior quantities are obtained for the exterior region:

$$
e^{-\lambda_e(r)} = 1 - \frac{2m}{r} + \frac{2m_{\Lambda e}(R_0, r)}{r},
\tag{6.42}
$$

where:

$$
m_{\Lambda e}(R_0, r) = -4\pi \int_{R_0}^{r} r'^2 \varepsilon_{\Lambda e}(r')\, dr',
\tag{6.43}
$$

m being some integration constant and R_0 a specific distance from the center (more details in Sect. 6.1.5). The exterior temporal metric coefficient is

$$
e^{\nu_e(r)} = e^{-\lambda_e(r)} e^{\frac{f_e(r)+C_e}{2}},
\tag{6.44}
$$

where C_e is used for the sake of continuity (see Sect. 6.1.5). The hydrostatic equilibrium equations for this region have now an extra term related to the non-isotropy (See also (3.53)):

$$
\frac{dp_{\Lambda re}}{dr} = -\frac{\left(\varepsilon_{\Lambda e}(r) + p_{\Lambda re}(r)\right)}{r\left[r - 2m + 2m_{\Lambda e}(r)\right]}
$$
$$
\times \left[m - m_{\Lambda e}(r) + 4\pi r^3 p_{\Lambda re}(r)\right] + \frac{2(p_{\Lambda\vartheta e} - p_{\Lambda re})}{r}.
\tag{6.45}
$$

Analogously to the interior case, complementary relations are needed in order to close the system of differential equations (6.43), (6.45). In the current case two relations are also needed. Different models can in principle be proposed as long as they satisfy two constrains: radial pressure continuity and strong fall-off dependence with distance. The former is related to our will of avoiding infinitesimally thin shells and the latter comes in order to satisfy standard Schwarzschild solutions at large enough distances. In the current study two models are considered:

- Model A: Continuous energy density at R_0.
  The following expression:

$$\varepsilon_{\Lambda e}(r) = \varepsilon_{\Lambda i R_0} \left(\frac{R_0}{r}\right)^5 + \frac{B}{8\pi} \frac{(r - R_0)}{r^6} \qquad (6.46)$$

where $\varepsilon_{\Lambda i R_0} = \varepsilon_{\Lambda i}(R_0)$ (units have been already expressed with $\kappa = c = 1$) and B is in arbitrary constant makes the Λ-term energy density to be continuous at R_0. Note, that the fall off in the radial distance is stronger than proposed earlier.

- Model B: Discontinuous energy density at R_0.
  Here the following profile is proposed:

$$\varepsilon_{\Lambda e} = \frac{B}{8\pi r^5}, \qquad (6.47)$$

where B is again an arbitrary constant. The fall off in the radial distance is here the same as proposed earlier.

Each of these possibilities will be complemented with a tangential pressure of the form:

$$p_{\Lambda \vartheta e}(r) = P_{\Lambda \vartheta R_0} \left(\frac{R_0}{r}\right)^5, \qquad (6.48)$$

where $P_{\Lambda \vartheta R_0} = p_{\Lambda \vartheta i}(R_0)$. Both energy density and tangential pressure do not have to be necessary continuous functions of the radius [5]. Within our study discontinuity is explored only for the former. The tangential pressure has been always considered continuous and with a similar dependence to that of the energy density.

6.1.5 Boundary Conditions

Discontinuous equations of state can lead in general to discontinuous metric coefficients and their first derivatives [27]. The presence of a rather regular matching between the different regions (instead of discontinuous radial pressure profiles which lead to infinitesimally thin shells) allows us to work with metric coefficients which are going to behave continuously at the boundaries. There are going to be two relevant distances: the first is where the baryonic pressure vanishes (denoted by R_0) and the second where the pc-metric practically does not differentiate itself from the standard Schwarzschild one (denoted by R_Λ, at this point the local effect of the Λ-term can be neglected). The radial metric coefficients can be determined by demanding from them to be equal at R_0. With (6.13), (6.42) m is restrained to be

$$m = M_m - M_{\Lambda i} \qquad (6.49)$$

with

$$M_m = m_m(R_0), \ M_{\Lambda i} = m_{\Lambda i}(R_0), \tag{6.50}$$

where $m_{\Lambda e}(R_0, R_0) = 0$ was used.

At sufficient large distance (bigger than R_Λ) it needs to behave Schwarzschild-alike, i.e. it should take the following form:

$$e^{-\lambda_e(r)} = 1 - \frac{2m'}{r} \quad r > R_\Lambda, \tag{6.51}$$

where m' has to be constant (whose value will depend on the model used). At these large enough distances the well known Schwarzschild relation $e^{\nu(R_\Lambda)} = -e^{-\lambda(R_\Lambda)}$ should be restored. Then, from (6.44):

$$C_e = -f_e(R_\Lambda). \tag{6.52}$$

With this value equating now between (6.44) and (6.33) at R_0:

$$C_i = -f_e(R_1) - f_i(R_0). \tag{6.53}$$

The value of m given by (6.49) was used as well as the property $f_e(R_0) = 0$. Using these values for C_i and C_e the temporal metric coefficients, from the interior and the exterior respectively, can be expressed as:

$$e^{\nu_i(r)} = e^{-\lambda_i(r)} e^{\frac{f_i(r) - f_e(R_1) - f_i(R_0)}{2}}, \tag{6.54}$$

$$e^{\nu_e(r)} = e^{-\lambda_e(r)} e^{\frac{f_e(r) - f_e(R_1)}{2}}. \tag{6.55}$$

6.1.6 Energy Conditions

The Einstein equations of GR can be in principle satisfied for a large number of energy-momentum tensors. The energy conditions (EC) precisely reduce their arbitrariness by making some "standard physics" demands for the energy-momentum tensor $T_{\mu\nu}$. Although the violation of these conditions does not mean the violation of energy conservation, it is always convenient to analyze to what extent they are satisfied. It follows a short summary with their content according to [28]:

- Weak Energy Condition (WEC):
  $T_{\mu\nu} t^\mu t^\nu \geq 0$ for all timelike vectors t^μ.
- Null Energy Condition (NEC):
  $T_{\mu\nu} l^\mu l^\nu \geq 0$ for all null vectors l^μ.

- Dominant Energy Condition (DEC):
  WEC: $T_{\mu\nu}t^{\mu}t^{\nu} \geq 0$ for all timelike vectors t^{μ}.
  $T^{\mu\nu}t_{\mu}$ non-spacelike vector.
- Null Dominant Energy Condition (NDEC):
  As the DEC but for null vectors instead of timelike ones.
- Strong Energy Condition (SEC):
  $T_{\mu\nu}t^{\mu}t^{\nu} \geq \frac{1}{2}T^{\lambda}_{\lambda}t^{\sigma}t_{\sigma}$ for all timelike vectors t^{μ}.

As time has passed different judgments about the EC have been formulated and some of them have been partially or totally abandoned [29]. In the current study only the first four will be analyzed. Depending on the properties of the energy-momentum tensor $T_{\mu\nu}$ the EC can be expressed as specific restrictions on the values of both the energy density and pressure (The EC will be proven in Exercise 6.3 for isotropic fluids and in 6.4 for anisotropic fluids.).

- Isotropic fluids:
  The energy-momentum tensor $T_{\mu\nu}$ for an isotropic fluid can be expressed as [28]:

$$T_{\mu\nu} = (\varepsilon + p)u_{\mu}u_{\nu} + pg_{\mu\nu} \tag{6.56}$$

where ε, p, u^{μ}, $g_{\mu\nu}$ stand for energy density, pressure, four-velocity and metric coefficients, respectively. In this case the EC take the following form [28]:

  – WEC:

$$\varepsilon \geq 0, \tag{6.57}$$

$$\varepsilon + p \geq 0. \tag{6.58}$$

  – NEC:
    Special case of the WEC where only (6.58) must be satisfied.
  – DEC:
    The WEC condition together with the additional demand of the vector $T^{\mu\nu}t_{\mu}$ being non-spacelike can be expressed as:

$$\varepsilon \geq |\, p\,|. \tag{6.59}$$

  – NDEC:
    In addition to the energy density and pressure values allowed by the DEC (6.59) this condition also allows negative energy densities as long as they satisfy $p = -\varepsilon$.

- Anisotropic fluids:
  When fluids are anisotropic their energy-momentum tensor $T_{\mu\nu}$ can be expressed as [30]:

$$T_{\mu\nu} = (\varepsilon + p_{\vartheta})u_{\mu}u_{\nu} + p_{\vartheta}g_{\mu\nu} + (p_r - p_{\vartheta})k_{\mu}k_{\nu}, \tag{6.60}$$

where ε, p_{ϑ}, p_r, u^{μ}, $g_{\mu\nu}$ denote energy density, tangential pressure, radial pressure, four-velocity and metric coefficients, respectively. The vector k^{μ} is an unitary space-like vector in the radial direction. In this case the EC take the following form [10]:

- WEC:

$$\varepsilon \geq 0, \tag{6.61}$$

$$\varepsilon + p_\vartheta \geq 0, \tag{6.62}$$

$$\varepsilon + p_r \geq 0. \tag{6.63}$$

- NEC:
Special case of the WEC where only (6.62), (6.63) must be satisfied.
- DEC:
The WEC condition together with the additional demand of the vector $T^{\mu\nu}t_\mu$ being non-spacelike can be expressed as:

$$\varepsilon \geq |p_\vartheta|, \tag{6.64}$$

$$\varepsilon \geq |p_r|. \tag{6.65}$$

- NDEC:
In addition to the energy density and pressure values allowed by the DEC (6.64), (6.65)) this condition also allows negative energy densities as long as they satisfy $p_\vartheta = -\varepsilon$ and $p_r = -\varepsilon$.

Exercise 6.3 (Energy conditions for an isotropic fluid)

Problem. Derive the relations between the pressures and densities for the WEC, NEC, DEC and NDEC, assuming an *isotropic fluid*.

Solution. The energy-momentum tensor is represented as:

$$T_{\mu\nu} = (\varepsilon + p)\, u_\mu u_\nu + p g_{\mu\nu}. \tag{6.66}$$

Let t^μ be an arbitrary time-like vector and let us express it as a linear combination of the four-velocity u^μ and an arbitrary null vector l^μ:

$$t^\mu = a u^\mu + b l^\mu. \tag{6.67}$$

Let us now see which values of a, b assure t^μ will remain being time-like:

$$g_{\mu\nu} t^\mu t^\nu < 0$$
$$g_{\mu\nu} (a u^\mu + b l^\mu)(a u^\nu + b l^\nu) < 0$$
$$g_{\mu\nu} (a^2 u^\mu u^\nu + 2ab u^\mu l^\nu + b^2 l^\mu l^\nu) < 0. \tag{6.68}$$

Taking into account that $g_{\mu\nu}u^\mu u^\nu = -1$ and $g_{\mu\nu}l^\mu l^\nu = 0$ the condition for the time-like character becomes

$$-a^2 + 2ab g_{\mu\nu}u^\mu l^\nu < 0$$
$$\text{or } a^2 - 2ab g_{\mu\nu}u^\mu l^\nu > 0, \tag{6.69}$$

an equation which will be of importance later on.

(a) WEC:

$$T_{\mu\nu}t^\mu t^\nu = T_{\mu\nu}\left[a^2 u^\mu u^\nu + 2ab u^\mu l^\nu + b^2 l^\mu l^\nu\right]$$
$$T_{\mu\nu}u^\mu u^\nu = (\varepsilon + p)\left(u_\mu u^\mu\right)^2 + p g_{\mu\nu}u^\mu u^\nu = \varepsilon$$
$$T_{\mu\nu}u^\mu l^\nu = (\varepsilon + p)\left(u_\mu u_\nu u^\mu l^\nu\right) + p g_{\mu\nu}u^\mu l^\nu = -\varepsilon u_\mu l^\mu$$
$$T_{\mu\nu}l^\mu l^\nu = (\varepsilon + p)\left(u_\mu u_\nu l^\mu l^\nu\right) + p g_{\mu\nu}l^\mu l^\nu = (\varepsilon + p)\left(u_\mu l^\mu\right)^2. \tag{6.70}$$

With the last three equations in (6.70) substituting into the first one in (6.70) gives and using (6.69):

$$\varepsilon\left[a^2 - 2ab g_{\mu\nu}u^\mu l^\nu\right] + (\varepsilon + p)\, b^2\left(u_\mu l^\mu\right)^2 \geq 0. \tag{6.71}$$

If we want (6.71) to be satisfied for all a and b, then:

$$\varepsilon \geq 0$$
$$\varepsilon + p \geq 0. \tag{6.72}$$

(b) NEC:

This condition is given by (6.70) and therefore is satisfied only by the second equation in (6.72).

(c) DEC:

Includes WEC, i.e. equations in (6.72) must be satisfied. In addition to that it also requires the vector $T^{\mu\nu}t_\mu$ to be non space-like, or equivalently:

$$T_{\mu\nu}T^\nu_\lambda t^\mu t^\lambda \leq 0$$
$$\left[(\varepsilon + p)\, u_\mu u_\nu + p g_{\mu\nu}\right]\left[(\varepsilon + p)\, u^\nu u_\lambda + p\delta^\nu_\lambda\right]t^\mu t^\lambda \leq 0$$
$$\left[(\varepsilon^2 - p^2)\, u_\mu u_\lambda + p^2 g_{\mu\nu}\right]t^\mu t^\lambda$$
$$= -(\varepsilon^2 - p^2)\left(u_\mu t^\mu\right)^2 + p^2 g_{\mu\lambda}t^\mu t^\lambda \leq 0. \tag{6.73}$$

If we want the last equation in (6.73) to be satisfied for all t^μ then (remember that $g_{\mu\lambda} t^\mu t^\lambda = t_\mu t^\mu < 0$):

$$|\varepsilon| \geq |p| . \tag{6.74}$$

Equations (6.74) and (6.72) can be summarized by:

$$\varepsilon \geq |p| . \tag{6.75}$$

Equivalently, this condition can be expressed as:

$$(\varepsilon + p)(\varepsilon - p) \geq 0 . \tag{6.76}$$

(d) NDEC:

It is like DEC but for null vectors. Therefore the second equation in (6.72) must be satisfied but densities may now be negative. From the second equation in (6.73), now with null vectors, we have

$$\left(\varepsilon^2 - p^2\right) \left(u_\mu l^\mu\right)^2 \leq 0, \tag{6.77}$$

which is then satisfied for all u^μ, l^ν if (6.74) (or (6.76)) is satisfied. The same values of pressure and energy density as in DEC are included plus those with negative energy density as long they satisfy $p = -\varepsilon$.

Exercise 6.4 (Energy conditions for an anisotropic fluid)

Problem. Derive the relations between the pressures and densities for the WEC, NEC, DEC and NDEC, using an *anisotropic fluid*.

Solution. The energy-momentum tensor is now given by:

$$T_{\mu\nu} = (\varepsilon + p_\vartheta) u_\mu u_\nu + p_\vartheta g_{\mu\nu} + (p_r - p_\vartheta) k_\mu k_\nu. \tag{6.78}$$

where $u^\mu u_\mu = -k^\nu k_\nu = -1$, $u_\eta k^\eta = 0$.

As you can check this tensor satisfies the diagonal structure.

Let us express again an arbitrary time-like vector t^μ according to (6.79),

$$t^\mu = au^\mu + bl^\mu, \tag{6.79}$$

and analyze then each one of the energy conditions:

(a) WEC:

$$T_{\mu\nu}t^{\mu}t^{\nu} = a^2 T_{\mu\nu}u^{\mu}u^{\nu} + 2ab T_{\mu\nu}u^{\mu}l^{\nu} + b^2 T_{\mu\nu}l^{\mu}l^{\nu}. \qquad (6.80)$$

Each of these terms gives the following results:

$$a^2 T_{\mu\nu}u^{\mu}u^{\nu} = a^2 \varepsilon$$
$$2ab T_{\mu\nu}u^{\mu}l^{\nu} = -2ab\varepsilon\,(u_{\nu}l^{\nu})$$
$$b^2 T_{\mu\nu}l^{\mu}l^{\nu} = b^2 \left[(\varepsilon + p_{\vartheta})\,(u_{\nu}l^{\nu})^2 + (p_r - p_{\vartheta})\,\left(k_{\mu}l^{\mu}\right)^2 \right]. \qquad (6.81)$$

Then the following relation is obtained:

$$\varepsilon\left[a^2 - 2ab\,(u_{\nu}l^{\nu})\right] + (\varepsilon + p_{\vartheta})\,b^2\,(u_{\nu}l^{\nu})^2 + (p_r - p_{\vartheta})\,b^2\,\left(k_{\mu}l^{\mu}\right)^2 \geq 0. \tag{6.82}$$

From where one could say the conditions for satisfying WEC are (remember from the former exercise that the factor $a^2 - 2ab\,(u_{\nu}l^{\nu})$ is positive):

$$\varepsilon \geq 0,$$
$$\varepsilon + p_{\vartheta} \geq 0,$$
$$p_r - p_{\vartheta} \geq 0. \qquad (6.83)$$

These relations are correct and they assure (6.82) is satisfied for all values of a, b. These relations can be converted to relations which are only between the energy density and the radial pressure (in principle one should never make such a conversion since the former relations are the ones which are directly obtained within this analysis). The argument is given by the inequalities: If $\varepsilon + p_{\vartheta} \geq 0$ then $p_{\vartheta} \geq -\varepsilon$. On the other hand if $p_r \geq p_{\vartheta}$, then $p_r \geq -\varepsilon$ will be also satisfied. Therefore (6.83) can be replaced by:

$$\varepsilon + p_r \geq 0. \qquad (6.84)$$

(b) NEC:
    This condition is given by the third equation in (6.81) which only needs the fulfillment of the last two equations in (6.83) or equivalently the second equation in (6.83) and (6.84).

(c) DEC:

The WEC condition has to be satisfied. Therefore (6.83) (or using instead of the third equation the (6.84)) must be satisfied. From now on we will refer to the last set of conditions. In addition to that the vector $T_{\mu\nu}t^\nu$ has to be non space-like:

$$T_{\mu\nu}T^\nu_\lambda t^\mu t^\lambda \leq 0$$
$$\left[(\varepsilon + p_\vartheta)\, u_\mu u_\nu + p_\vartheta g_{\mu\nu} + (p_r - p_\vartheta)\, k_\mu k_\nu\right]$$
$$\times \left[(\varepsilon + p_\vartheta)\, u^\nu u_\lambda + p_\vartheta \delta^\nu_\lambda + (p_r - p_\vartheta)\, k^\nu k_\lambda\right] t^\mu t^\lambda \leq 0. \qquad (6.85)$$

Doing some algebraical work the former expression can be expressed as:

$$- \left(\varepsilon^2 - p_\vartheta^2\right) \left(u_\mu t^\mu\right)^2 - \left(p_\vartheta^2 - p_r^2\right) (k_\nu t^\nu)^2 + p_\vartheta^2 g_{\mu\nu} t^\mu t^\lambda \leq 0. \qquad (6.86)$$

which is satisfied for all vectors u^μ, v^μ, t^μ if:

$$\varepsilon^2 - p_\vartheta^2 \geq 0,$$
$$p_\vartheta^2 - p_r^2 \geq 0. \qquad (6.87)$$

Following analogous arguments, the former expression can be replaced by

$$\varepsilon^2 - p_r^2 \geq 0. \qquad (6.88)$$

These expression can be expressed also as:

$$|\varepsilon| \geq |p_r|$$
$$|\varepsilon| \geq |p_\vartheta|. \qquad (6.89)$$

Another way to express them is:

$$(\varepsilon + p_\vartheta)\,(\varepsilon - p_\vartheta) \geq 0,$$
$$(\varepsilon + p_r)\,(\varepsilon - p_r) \geq 0. \qquad (6.90)$$

(d) NDEC:

Placing null vectors in the first equation in (6.85) the following relation is obtained:

$$- \left(\varepsilon^2 - p_\vartheta^2\right) \left(u_\mu l^\mu\right)^2 - \left(p_\vartheta^2 - p_r^2\right) (k_\nu l^\nu)^2 \leq 0, \qquad (6.91)$$

which is satisfied for all vectors if (6.89) is fulfilled. Since now (6.83) is not demanded, negative values of the energy density are allowed as long as they satisfy $p_\vartheta = -\varepsilon$ and $p_r = -\varepsilon$.

6.2 Numerical Framework, Results and Discussions

The system of four differential equations (6.14), (6.15), (6.35) and (6.36) complemented with two equations of state have been numerically integrated using a 4th-order Runge-Kutta algorithm which solves the system for a given value of the central baryonic energy density ε_{mc} and the central Λ-term pressure $p_{\Lambda c}$. The system is solved initially for the values of r, p_m, m_m, m_Λ at the center and then these values are used to compute for the next radial step. This procedure is done until some boundary condition is reached, in our case the criteria of vanishing baryonic pressure p_m has been employed. The number of iterations N is directly related to the length of the radial step δr, convergence of solutions has been checked for different values of these parameters.

Pressure and energy density have been treated in km^{-2} while distance and masses in km. In order to compare with GR final results have been converted into standard units, i.e. $\mathrm{MeV \cdot fm}^{-3}$ and solar masses $M_\odot$. Since the same units for both pressure and energy density have been employed during calculations, multiples of the nuclear energy density $\varepsilon_0 = 141 \; \mathrm{MeV \cdot fm}^{-3}$ have been used as measure of the given values of ε_{mc} and $p_{\Lambda c}$.

Exercise 6.5 (Schwarzschild radius of the sun)

Calculate the mass of the sun in terms of length (km). What is the value of the Schwarzschild radius of the sun?

Solution. The mass of the sun, in terms of km, is given by

$$m = \frac{\kappa}{c^2} M, \qquad (6.92)$$

with κ being the gravitational constant, c the light velocity and M the mass of the sun in kg.

The basis values, available in any book on units, are

$$\kappa = 6.67384 \times 10^{-11} \; \frac{\mathrm{m}^3}{\mathrm{kg \; s}^2},$$

$$c = 2.99792458 \times 10^8 \; \frac{\mathrm{m}}{\mathrm{s}},$$

$$M_S = 1.98855 \times 10^{30} \; \mathrm{kg}, \qquad (6.93)$$

where M_S is the mass of the sun.

Substituting these values into (6.92) gives

$$m_S = 1.476627323 \times 10^3 \; \mathrm{m} = 1{,}476627323 \; \mathrm{km}, \qquad (6.94)$$

i.e., the Schwarzschild radius of the sun is 2m, i.e., about 3 km.

Exercise 6.6 (Units)

Problem. Suppose, you use the natural units

$$\kappa = 1 \text{ and } c = 1. \tag{6.95}$$

What are the units of time, mass, density and energy in these units?

Solution. Using the list of values for the gravitational constant and velocity of light, we have

$$\kappa = 6.67384 \times 10^{-11} \frac{m^3}{kg\ s^2} = 1,$$
$$c = 2.99792458 \times 10^8 \frac{m}{s}. \tag{6.96}$$

From the second equation we get

$$1\,s = 2.99792458 \times 10^8\ m. \tag{6.97}$$

This implies that time is measured in length!

In order to get the units of the mass, we part from the first equation in (6.96) and (6.97)

$$1\,kg = 6.67384 \times 10^{-11} \frac{m^3}{s^2}$$
$$= \frac{6.67384}{(2.99792458)^2} \times 10^{-11-16}\ m$$
$$= 0.74256 \times 10^{-27}\ m, \tag{6.98}$$

where we made use of the new units of c.

Multiplying this value with the mass of the sun, gives just $m \approx 1.5$ km, half of the Schwarzschild radius, which confirms the result of Exercise 6.5.

The units of energy are $1\,J = 6.2422 \times 10^{13}$ MeV. One J is given by

$$1 \frac{kg\ m^2}{s^2} = 8.2622 \times 10^{-45}\ m, \tag{6.99}$$

where we made use of the new units of time and kg, derived above. Equating this with one J in units of MeV gives

$$1\,MeV = 1.3236 \times 10^{-58}\ m. \tag{6.100}$$

The energy density is given in units of MeV/fm^3. One fm is 10^{-15} m. One MeV can also be given in meters, according to (6.100). The nuclear density is known to be

$$\varepsilon_0 = 141 \, \frac{\text{MeV}}{\text{fm}^3}. \tag{6.101}$$

Converting in (6.101) the fm in m and using (6.100) leads to

$$\varepsilon_0 = 141 \frac{1.3236 \times 10^{-58}}{10^{-45}} \frac{1}{\text{m}^2}$$

$$= 1.866276 \times 10^{-11} \, \frac{1}{\text{m}^2}$$

$$= 1.866276 \times 10^{-5} \, \frac{1}{\text{km}^2}. \tag{6.102}$$

The same procedure has been carried out in order to solve the system of differential equations corresponding to the exterior shell.

The Λ-term, which is contained in pc-GR equations in a natural way, may be physically interpreted as the average contribution of vacuum fluctuations. As pointed out before, a semi-classical study of this phenomena has been carried out on a Schwarzschild background having, for specific types of vacuum, energy densities with strong falling-off terms as $1/r^6$ [10–13]. In [14] speculations have been made about this contribution being responsible for halting gravitational collapses under certain conditions. Baryonic properties are expected to change once the Λ-term is included within the equations. Understanding these changes is very important since, given the local character of vacuum fluctuations, it may be only the baryonic component what is measured at large distances after all. Under the same conditions, calculations for standard GR were carried out in order to compare between both theories. Since only one model has been studied for the interior its results are going to be shown first. Figures 6.2 and 6.3 show the set of results for a single star which has a fixed value of the baryonic central energy density, $\varepsilon_{mc} = 5\varepsilon_0$, but different values of the coefficient α.

Figure 6.2 shows the energy density profile ε_m for different values of α. For a fixed radius, higher ε_m are obtained as $|\alpha|$ increases within the selected range, showing that increasing the accumulation of the Λ-term results in a higher baryon compactness. This causes the stars radii to increase, i.e. bigger stars.

This effect can be also interpreted in terms of pressure profiles. Figure 6.3 shows how the baryonic pressure increases as $|\alpha|$ increases within the selected range. The surface of the star was defined to be located at the radial value for which the baryonic pressure vanishes.

Remarkably, high baryonic masses can be predicted within this model. Calculations for a family of stars with central baryonic energy densities within the range $[1–10]\,\varepsilon_0$ have also been made. These results help us to gain a wider insight about the

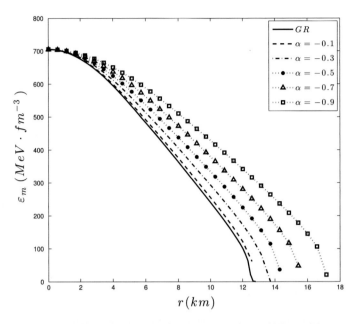

Fig. 6.2 Baryonic energy density profile for different values of the coefficient α. The central Λ-term pressure $p_{\Lambda c}$ has been fixed to $1\varepsilon_0$

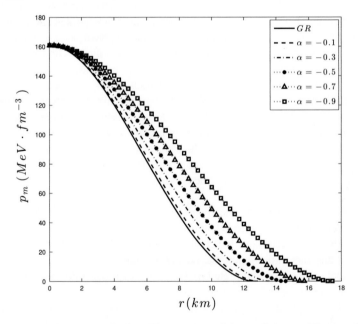

Fig. 6.3 Baryonic pressure profile for different values of the coefficient α. The central Λ-term pressure $p_{\Lambda c}$ has been fixed to $1\varepsilon_0$

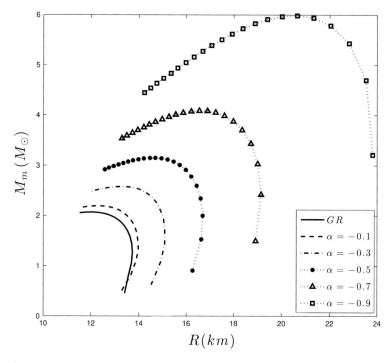

Fig. 6.4 Baryonic total mass versus total radius for different values of the coefficient α. The central Λ-term pressure $p_{\Lambda c}$ has been fixed to $1\varepsilon_0$

properties of these compact objects. Figure 6.4 shows the total baryonic mass of the stars as function of the total radius, within this range keeping constant the Λ-term central pressure at $p_{\Lambda c} = 1\varepsilon_0$ while α is changed in the same range as before. It can be seen how within this model configurations which may be stable can be achieved up to $6M_\odot$ when the highest value of $|\alpha|$ is used.

In Fig. 6.5 the total baryonic mass is again plotted, but this time as a function of the central baryonic energy density. It can be observed the necessary stability condition $\frac{dM_m}{d\varepsilon_{mc}} > 0$ being always satisfied at some range. As $|\alpha|$ is increased the former range gets narrower expressing that for a specific central baryonic energy density there are fewer compact objects which may be stable as more baryons are pulled together. *Nevertheless, obtaining larger baryonic masses is not possible within this linear coupling assumption.*

Other models need to be employed, but from a classical point of view *there are no obvious arguments in order to improve such a selection.*

The baryonic compactness is shown in Fig. 6.6, where it can be appreciated how the high-mass objects present a compactness close to the unity. This may in principle allow, as long as certain conditions are satisfied in the exterior region, *to mimic the properties of black holes.* As previously mentioned two models were studied for the non-isotropic exterior shell. The tangential pressure was kept continuous at the

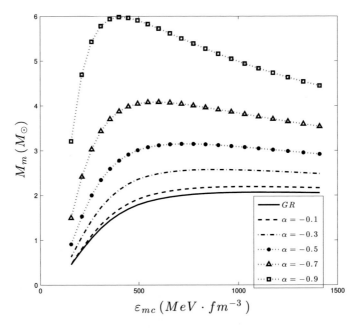

Fig. 6.5 Baryonic total mass versus central baryonic energy density for different values of the coefficient α. The central Λ-term pressure $p_{\Lambda c}$ has been fixed to $1\varepsilon_0$

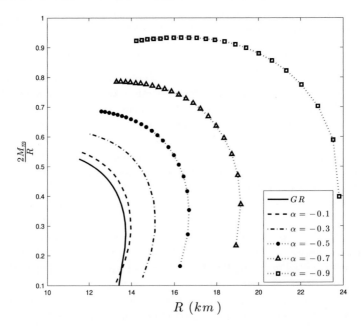

Fig. 6.6 Baryonic compactness versus total radius for different values of the coefficient α. The central Λ-term pressure $p_{\Lambda c}$ has been fixed to $1\varepsilon_0$

boundary $r = R_0$ following the power law given by (6.48), while the energy density obeyed the following expressions

- Model A: Continuous energy density at R_0.
  Placing (6.46), (6.49) into (6.42) the following expression for the radial exterior metric coefficient is obtained:

$$
e^{-\lambda_e(r)} = 1 - \frac{2}{r}\left(M_m - M_\Lambda + 2\pi \varepsilon_{\Lambda i\, R_0} R_0^3 + \frac{B}{12 R_0^2} \right)
$$
$$
- \frac{B R_0}{3 r^4} + \frac{8\pi \varepsilon_{\Lambda i\, R_0} R_0^5 + B}{2 r^3}. \tag{6.103}
$$

The constant B has been so far just an arbitrary constant. By simple inspection of (6.103) one realizes that picking B to be:

$$
B_m = 12 M_{\Lambda i} R_0^2 - 24\pi \varepsilon_{\Lambda i\, R_0} R_0^5. \tag{6.104}
$$

Equation (6.103) transforms to:

$$
e^{-\lambda_e(r)} = 1 - \frac{2 M_m}{r} - \frac{4 M_{\Lambda i} R_0^3 - 8\pi \varepsilon_{\Lambda i\, R_0} R_0^6}{r^4}
$$
$$
+ \frac{6 M_{\Lambda i} R_0^2 - 8\pi \varepsilon_{\Lambda i\, R_0} R_0^5}{r^3}. \tag{6.105}
$$

Giving the possibility that at large enough distances only the baryonic component will make a contribution behaving Schwarzschild alike (i.e. proportional to the inverse of the distance). Figure 6.7 (and its zoom for a section of the exterior region Fig. 6.8) show how the Λ-term energy density looks like. There is a continuous matching with the interior (black), a relative small cusp and then a strong fall-off takes place.

The tangential Λ-term pressure has a similar form (Fig. 6.9) only with a stronger fall-off character. The difference between them can be better appreciated in Fig. 6.9 where both pressures have been plotted in a single plot. Figure 6.10 shows the radial metric coefficient.

- Model B: Discontinuous energy density at R_0.
  Placing now (6.47), (6.49) into (6.42), the following expression for the radial exterior metric coefficient is obtained:

$$
e^{-\lambda_e(r)} = 1 - \frac{2\left(M_m - M_{\Lambda i} + \frac{B}{4 R_0^2} \right)}{r} + \frac{B}{2 r^3}. \tag{6.106}
$$

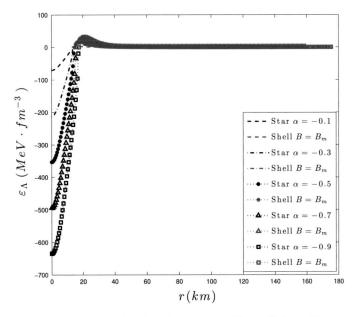

Fig. 6.7 Λ-term energy density profiles for different values of the coefficient α. The central Λ-term pressure $p_{\Lambda c}$ has been fixed to $1\varepsilon_0$. The parameter B_m has been chosen (Model A)

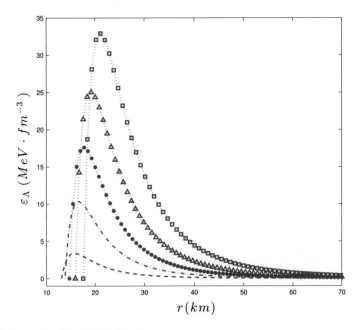

Fig. 6.8 Zoom from Fig. 6.7 (Model A)

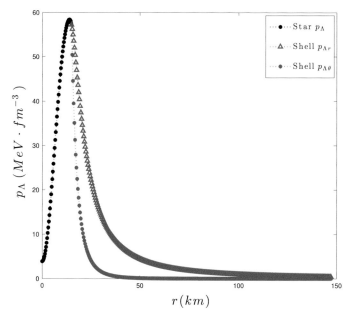

Fig. 6.9 Λ-term radial and tangential pressure profiles for $\alpha = -0.5$. The central Λ-term pressure $p_{\Lambda c}$ has been fixed to $1\varepsilon_0$. The parameter B_m has been chosen (Model A)

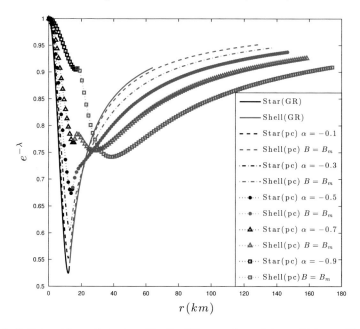

Fig. 6.10 Radial metric coefficients profiles for different values of the coefficient α. The central Λ-term pressure $p_{\Lambda c}$ has been fixed to $1\varepsilon_0$. The parameter B_m has been chosen (Model A)

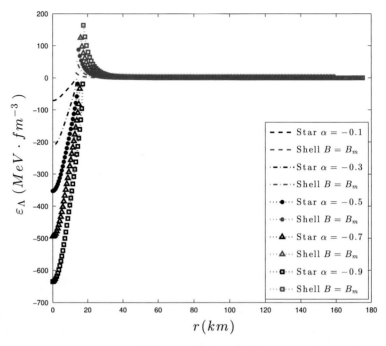

Fig. 6.11 Λ-term energy density profiles for different values of the coefficient α. The central Λ-term pressure $p_{\Lambda c}$ has been fixed to $1\varepsilon_0$. The parameter B_m has been chosen (Model B)

If B is chosen now to be $B_m = 4M_{\Lambda i} R_0^2$, Eq. (6.106) becomes:

$$e^{-\lambda_e(r)} = 1 - \frac{2M_m}{r} + \frac{2M_{\Lambda i} R_0^2}{r^3}. \tag{6.107}$$

Giving again the possibility of having only baryonic contributions at large enough distances. Figure 6.11 shows the analogous quantities to those previously shown to the continuous case.

Despite they behave quite similar, there is a discontinuity now on the derivative of the temporal coefficients coming as a consequence of the equation of state discontinuity. Finally, in (Fig. 6.12) the pressure p_Λ of the dark energy is depicted versus the radial distance r. Having used an anisotropic fluid outside the star, produces a different behavior for the radial to the tangential pressure, the latter being consistently smaller.

6.3 Resuming the Results Presented in This Chapter

The pc-GR theory includes itself a Λ-term which is believed to represent the vacuum fluctuations contribution. Within this frame a study of neutron stars has been made describing the mentioned term as an additional energy-momentum tensor.

The interior region was characterized by having a linear coupling between the energy densities of both components. The main purposes of such a model were:

- A negative dark interior energy density.
- A positive total interior energy density.
- Simplicity.

Within this approach, the baryonic mass of the neutron stars can be raised to values of almost $6M_\odot$ without having the formation of an event horizon. Higher baryonic masses could in principle be obtained but a more realistic model for the coupling of the components has to be employed. This cannot be achieved from a purely classical point of view.

Since pc-GR includes the mentioned extra-term, even where there is no other form of matter, there is going to be a surrounding Λ-term shell which has been studied under two possible models with continuous (model A) and discontinuous (model B)

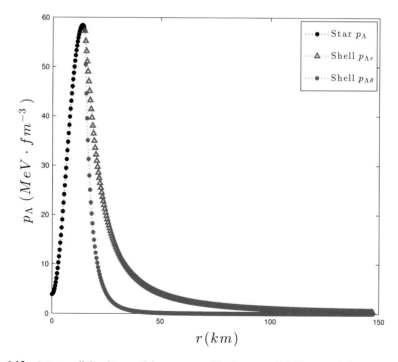

Fig. 6.12 Λ-term radial and tangential pressure profiles for $\alpha = -0.5$. The central Λ-term pressure $p_{\Lambda c}$ has been fixed to $1\varepsilon_0$. The parameter B_m has been chosen (Model B)

energy density at the boundary $r = R_0$. Discontinuities at the derivative of the temporal metric coefficient on model B has driven us to conclude that a continuous energy density profile is needed, therefore model A is preferable.

A particularly noteworthy feature is the possibility of letting the baryonic mass be the only one to influence at enough large distances. As discussed in Sect. 6.2 this can be achieved by picking a particular value of the constant B denoted by B_m. As a result the Λ-term component influence will strongly decay with distance making the differences with respect to Schwarzschild noticeable only for close enough observers.

The baryonic compactness increases as $|\alpha|$ increases and it approaches the one of standard GR black holes for high baryonic masses, i.e. these massive neutron stars have their Schwarzschild radii just under their surface. To a distant observer they will resemble a black hole since they may be highly redshifted but unlike the former their redshift should remain finite.

References

1. M.A. Abramowicz, W. Klu, No observational proof of the black-hole event horizon. A&A **396**, L31 (2002)
2. G. Chapline, E. Hohlfeld, R.B. Laughlin, D.I. Santiago, Quantum phase transitions and the breakdown of classical general relativity. Philos. Mag. Part B **81**, 235 (2001)
3. P. Mazur, E. Mottola, *Gravitational Condensate Stars: an Alternative to Black Holes*. arXiv:gr-qc/0109035 (2001)
4. M. Visser, D.L. Wiltshire, Stable gravastars, an alternative to black holes. Class. Quantum Gravity **21** (2004)
5. C. Cattoen, T. Faber, M. Visser, Gravastars must have anisotropic pressure. Class. Quantum Gravity **2**, 4189 (2005)
6. I. Dymnikova, Vacuum nonsingular black hole. Gen. Relativ. Gravit. **24**, 235 (1992)
7. I. Dymnikova, The algebraic structure of a cosmological term in spherical symmetric solutions. Phys. Lett. B **472**, 33 (2000)
8. F.S.N. Lobo, Stable dark energy stars. Class. Quantum Gravity **23**, 1525 (2006)
9. C.R. Ghezzi, Anisotropic dark energy stars. Astrophys. Space Sci. **333**, 437 (2011)
10. M. Visser, Gravitational vacuum polarization I: energy conditions in the Hartle-Hawking vacuum. Phys. Rev. D **54**, 5103 (1996)
11. M. Visser, Gravitational vacuum polarization II: energy conditions in the Bouleware vacuum. Phys. Rev. D **54**, 5116 (1996)
12. M. Visser, Gravitational vacuum polarization III: energy conditions in the (1+)-dimensional Schawrzschild spacetime. Phys. Rev. D **54**, 5123 (1996)
13. M. Visser, Gravitational vacuum polarization IV: energy conditions in the Unruh vacuum. Phys. Rev. D **56**, 936 (1997)
14. C. Barceló, S. Liberati, S. Sonego, M. Visser, Fate of gravitational collapse in semiclassical gravity. Phys. Rev. D **77**, 044032 (2008)
15. I. Rodríguez, P.O. Hess, S. Schramm, W. Greiner, Neutron stars within pseudo-complex general relativity. J. Phys. G **41**, 105201 (2014)
16. R. Adler, M. Bazin, M. Schiffer, *Introduction to General Relativity*, 2nd edn. (McGraw Hill, New York, 1975)
17. J. Oppenheimer, G. Volkov, On massive neutron cores. Phys. Rev. Online Arch. **55**, 374 (1939)
18. R.C. Tolman, Static solutions of Einstein's field equations for spheres of fluid. Phys. Rev. **55**, 364 (1939)

19. V. Dexheimer, S. Schramm, Proto-neutron and neutron stars in a chiral SU(3) model. Astrophys. J. **683**, 943 (2008)
20. P. Papazoglou, S. Schramm, J. Schaffner-Bielich, H. Stöcker, W. Greiner, Chiral Lagrangian for strange hadronic matter. Phys. Rev. C **57**, 2576 (1998)
21. P. Papazoglou, D. Zschiesche, S. Schramm, J. Schaffner-Bielich, H. Stöcker, W. Greiner, Nuclei in a chiral SU(3) model. Phys. Rev. C **59**, 411 (1999)
22. S. Schramm, Deformed nuclei in a chiral model. Phys. Rev. C **66**, 064310 (2002)
23. J. Antoniadis, P.C.C. Freire, N. Wex et al., A massive pulsar in a compact relativistic binary. Science **340**, 448 (2013)
24. P.B. Demorest, T. Pennucci, S.M. Ramos, M.S.E. Roberts, J.W.T. Hessels, A two-solar mass neutron star measured using Shapiro delay. Nature **467**, 1081 (2010)
25. W. Israel, Singular hypersurfaces and thin shells in general relativity II. Nuovo Cim. B **571**, 10 (1931)
26. F.S.N. Lobo, P. Crawford, Stability analysis of dynamic thin shells. Class. Quantum Gravity **22**, 1 (2005)
27. P. Mazur, E. Mottala, Gravitational condensate stars: an alternative to black holes. PNAS **101**, 9545 (2004)
28. S. Carroll, *Spacetime and Geometry*, An introduction to general relativity (Addison-Wesley, San Francisco, 2004)
29. M. Visser, C. Barceló. *Energy Conditions and their Cosmological Implications.* arXiv:gr-qc/0001099 (2000)
30. M.H. Daouda, M.E. Rodrigues, M. Houndjo, Anisotropic fluid for a set of non-diagonal tetrads in gravity. Phys. Lett B **715**(241), (2012)

Chapter 7
Pseudo-complex Differential Geometry

General Relativity is a theory of gravitation, with the metric as the dynamical field describing the curvature of space-time itself. This is different to other field theories, where one considers the propagation of a field on a predetermined, independently given space-time. In this sense, General Relativity is a theory very much connected with geometry, although it should not be considered only as an application of pseudo-Riemannian geometry to physical space and time [1]. One important symmetry principle of General Relativity is that of general covariance, according to which the physics should be independent of its specific coordinate representation. Such a coordinate-independent formulation is conducted using tools from pseudo-Riemannian differential geometry. In the literature one often observes two, not necessarily mutually exclusive ways of formulating the theory. In the more classical algebraic approach one considers coordinates, vector components, etc. with respect to some specific coordinate system, but emphasizes the transformation properties of these quantities with respect to a coordinate transformation [1, 2]. In this way one assures to follow the principle of general covariance, but is able to keep the language on a less abstract and formal level suitable to physical applications. The second, modern geometric approach uses a more formal language, describing points in the space-time, vectors, etc. independently from specific coordinate systems as abstract mathematical entities. In this formulation general covariance is inherently ensured. It should be emphasized, that both approaches describe the same physics, but in a slightly different language. Despite the price of using mathematically elaborate concepts, the modern language of differential geometry has the advantage of an often more elegant formulation, with its application not restricted to General Relativity, but of importance to other fields as well, especially high energy physics [3–5].

© Springer International Publishing Switzerland 2016
P.O. Hess et al., *Pseudo-Complex General Relativity*,
FIAS Interdisciplinary Science Series, DOI 10.1007/978-3-319-25061-8_7

7.1 A Short Introduction to Differential Geometry

7.1.1 Topology

The fundamental mathematical structure we are interested in is a *differential manifold*, which is a *topological space* with some specific properties which allows to define coordinates of points in this space in a suitable way. As a preliminary work we have to introduce the concept of a topological space. These are very general mathematical structures, of which differential manifolds are only a subset. That is, every manifold is a topological space, but not every topological space is a manifold [5].

A topological space consists of some set M together with a collection of subsets $\mathcal{T}$, which are often called the *open sets*. This pair $(M, \mathcal{T})$ is a topological space if it fulfills the following conditions:

- The empty set $\emptyset$ and the full set M itself are members of the collection $\mathcal{T}$.
- The union of any number of subsets which are elements of $\mathcal{T}$ is again a member of $\mathcal{T}$.
- The intersection of a finite number of subsets which are elements of $\mathcal{T}$ is again a member of $\mathcal{T}$.

We want to make this concept clear by considering open sets in $\mathbb{R}^2$. The Euclidean distance between two points (x_1, x_2) and $(y_1, y_2) \in \mathbb{R}^2$ is given by

$$d(x, y) = \sqrt{(y_1 - x_1)^2 + (y_2 - x_2)^2}. \tag{7.1}$$

The open ball $U_\varepsilon(x)$ around a point x is determined by all points in $\mathbb{R}^2$ which have a distance less than ε to x:

$$U_\varepsilon(x) = \left\{ y \in \mathbb{R}^m : d(x, y) < \varepsilon \right\}. \tag{7.2}$$

We now define a set U to be open if for every $x \in U$ there is an open ball $U_\varepsilon(x)$ around x, which is entirely contained in U (that is $U_\varepsilon(x) \subset U$). From this definition it is easy to see, that $\mathbb{R}^2$ itself is open, whereas we technically have to define the empty set $\emptyset$ to be open as well. If we now denote the collection of all such open sets as $\mathcal{T}$, then the pair $(\mathbb{R}^2, \mathcal{T})$ is a topological space. Note that this procedure is still valid for $\mathbb{R}^m$ with arbitrary dimension m. This example also illustrates, why in the above definition we have to demand *finite* intersections only. If we would consider the intersection of infinitely many subsequently shrinking open balls around some point x, we ultimately would obtain only the point x itself, which does not represent an open set. This does not happen for a *finite* number of intersections of such open balls.

It should be emphasized, that for a given set M in general several topologies are possible. For instance one can always choose the *trivial* topology, where only the set M and the empty set $\emptyset$ are defined to be open, that is $\mathscr{T} = \{\emptyset, M\}$, or the *discrete* topology, where $\mathscr{T}$ contains all possible subsets of M.

For two topological spaces M, M', we can define maps $f : M \rightarrow M'$. Such a map is called *continuous*, if the inverse image of an open set in M' is an open set in M. It should be emphasized, that the inverse statement does not have to be true. Consider for example the open interval $(-1, 1) \in \mathbb{R}$, which under the continuous map $f(x) = x^2$ is mapped to the interval $[0, 1)$, which is not open due to the closed lower boundary. However, we will mostly consider *homeomorphisms*, which are continuous maps having a continuous inverse. Such a homeomorphism $f : M \rightarrow M'$ bijectively maps open sets from M to open sets in M' and vice versa.

7.1.2 Differential Manifolds

In the last section we presented the general definition of a topological space and showed how this concept can be realized by the well-known open sets in $\mathbb{R}^m$. We also introduced homeomorphisms, which map open sets bijectively from one topological space to another. We now use these concepts to define a differential manifold, which is basically a topologically space together with a set of homeomorphisms providing points of the given topological space with coordinates in $\mathbb{R}^m$. Consider some open set U of a topological space M, that is $U \subset M$. Assume there is a homeomorphism $\phi : U \rightarrow \mathbb{R}^m$. This homeomorphism maps the open set U bijectively to an open set $\phi(U)$ in $\mathbb{R}^m$. We thus have for every point $p \in U$ exactly one point $x \in \mathbb{R}^m$, or

$$\phi(p) = \left\{x^1(p), \dots, x^m(p)\right\} = x^\mu(p). \tag{7.3}$$

The open set U is called the *coordinate neighborhood*, while the homeomorphism ϕ as well as the set $x^\mu(p)$ is called the *coordinate (function)*. The coordinate neighborhood together with the coordinate function is often called a *chart* or a *coordinate system*. Using this chart, we thus have identified every point p of the coordinate neighborhood, which lies in the abstract topological space, with a coordinate $x^\mu(p)$, which lies in the well-known space $\mathbb{R}^m$. The next step is to assure that not only the points in this specific coordinate neighborhood U are provided with a coordinate, but all points of the topological space M. For this purpose we assume a collection of charts $\{(U_i, \phi_i)\}$, such that the respective coordinate neighborhoods cover the entire space M:

$$M = \bigcup_i U_i. \tag{7.4}$$

Such a collection of charts is called an *atlas*. This atlas provides for every point $p \in M$ at least one coordinate $\phi(p)$ according to some chart (U, ϕ). But since the coordinate neighborhoods have to overlap in order to cover the entire manifold, often a point p will be part of two or more coordinate neighborhoods. We thus have to take care of *coordinate transformations*. Consider two charts (U_i, ϕ_i) and (U_j, ϕ_j) with overlapping coordinate neighborhoods:

$$U_{ij} = U_i \cap U_j \neq \emptyset. \tag{7.5}$$

Since the overlap U_{ij} is contained in both charts, we have two different coordinates for all points $p \in U_{ij}$:

$$\phi_i(p) = x^\mu, \quad \phi_j(p) = y^\nu. \tag{7.6}$$

We now can connect these coordinates by the map

$$\Psi_{ij} = \phi_i \circ \phi_j^{-1}, \tag{7.7}$$

which takes a coordinate $y^\nu(p)$, maps it to the corresponding point $p \in M$, and then maps this point to the other coordinate $x^\mu(p)$. This yields a *coordinate transformation* $x^\mu(y^\nu) = [\Psi_{ij}(y^\nu)]^\mu$, with Ψ_{ij} as a map $\mathbb{R}^m \to \mathbb{R}^m$. We demand that these coordinate transformations are infinitely differentiable in the sense well-known from real analysis [5].

This approach gives us the definition of an m-dimensional *differentiable manifold* as a topological space M, together with an atlas of charts $\{(U_i, \phi_i)\}$ which provide coordinates $\phi_i(p) = x^\mu(p)$ for each point $p \in U$, respectively, and whose coordinate transformations $\Psi_{ij} : \mathbb{R}^m \to \mathbb{R}^m$ are infinitely differentiable.

Note that referring to the formulation of General Relativity, an algebraic approach as mentioned in the introductory remarks at the beginning of this chapter, would mainly refer to the coordinates x^μ, y^ν and the respective coordinate transformations. An approach based on differential geometry rather puts the focus on the abstract point p as an element of the manifold M, and considers the coordinate $x^\mu(p)$ merely as a coordinate of this point obtained using some specific chart.

7.1.3 Vectors and Tensors

After presenting the mathematical structure of a differentiable manifold, we now want to introduce some basic objects.

A *function* f is defined as a map from a manifold M into the real numbers $\mathbb{R}$, that is

$$f : M \to \mathbb{R}, \quad p \mapsto f(p). \tag{7.8}$$

Note that this function is defined on the abstract space M, but for a given coordinate system (U, ϕ) it can be written in terms of the respective coordinates:

$$f \circ \phi^{-1} : \mathbb{R}^m \to \mathbb{R}, \quad x^\mu \mapsto f[\phi^{-1}(x^\mu)] = f(x^\mu). \tag{7.9}$$

Recall that the homeomorphism $\phi(p)$ associates to every point $p \in U$ a coordinate $x^\mu(p)$, so the inverse $\phi^{-1}(x^\mu)$ yields the abstract point $p \in U$ corresponding to the coordinate x^μ. The coordinate representation of the function f then maps the coordinate x^μ to a real value $f(x^\mu)$. Note that the same function f in general has different representations for different coordinate systems (U_i, ϕ_i) [5].

The complimentary object to a function is a *curve* $c(\lambda)$, which is a map from an interval (a, b) in $\mathbb{R}$ into the manifold M:

$$c: (a, b) \to M, \quad \lambda \mapsto c(\lambda). \tag{7.10}$$

Given a coordinate system (U, ϕ), such a curve has a coordinate representation $x^\mu(\lambda)$:

$$\lambda \mapsto x^\mu(\lambda) = (\phi \circ c)(\lambda), \tag{7.11}$$

that is, a curve in the abstract space M is mapped to the coordinate space $\mathbb{R}^m$.

An obvious next step is to consider the composition $(f \circ c)(\lambda)$, that is the value of a function on a curve:

$$f \circ c: (a, b) \to \mathbb{R}, \quad \lambda \mapsto f[c(\lambda)]. \tag{7.12}$$

We can calculate the change of f along the curve:

$$\frac{d f[c(\lambda)]}{d\lambda}. \tag{7.13}$$

Note that this expression does not depend on any particular coordinate system. Using the chain rule, we can associate this general expression with the coordinate representations of the function f and the curve $c(\lambda)$:

$$\frac{d f[c(\lambda)]}{d\lambda} = \frac{\partial f(x^\mu)}{\partial x^\mu} \frac{dx^\mu(\lambda)}{d\lambda}. \tag{7.14}$$

One observes that this expression is a combination of two contributions:

- The change of the function $f(x^\mu)$ with respect to the coordinates x^μ (differential of f).
- The change of the coordinate representation of the curve $x^\mu(\lambda)$ with respect to the curve parameter λ (tangent vector to the curve c).

We now consider both contributions separately as operators. The *tangent vector* to a curve $x^\mu(\lambda)$ is defined as the operator

$$\frac{d}{d\lambda} = \frac{dx^\mu(\lambda)}{d\lambda} \frac{\partial}{\partial x^\mu}, \tag{7.15}$$

which acts on a function f according to

$$\frac{d}{d\lambda}[f] = \frac{dx^\mu(\lambda)}{d\lambda}\frac{\partial}{\partial x^\mu}[f] = \frac{dx^\mu(\lambda)}{d\lambda}\frac{\partial f(x^\mu)}{\partial x^\mu}, \tag{7.16}$$

where we assume that these expressions are evaluated at one specific point $p = c(\lambda_0)$ of the curve. It can be shown that the set of all these operators forms a vector space, called the *tangent space* $T_p M$ at the point p, with the set $\{\partial/\partial x^\mu\}$ called the *coordinate basis* of this vector space. Any tangent vector then can be written as

$$X = X^\mu \frac{\partial}{\partial x^\mu} \tag{7.17}$$

with the tangent vector to a curve $x^\mu(\lambda)$ having components

$$X^\mu = \frac{dx^\mu(\lambda)}{d\lambda}. \tag{7.18}$$

It should be emphasized, that the tangent vector $X \in T_p M$ is a mathematical object which exists independently from any coordinate system. If we consider two coordinate systems with coordinates x^μ and $x^{\mu'}$, respectively, we have two different sets of coordinate basis vectors and vector components, both representing the same abstract vector X:

$$X = X^\mu \frac{\partial}{\partial x^\mu} = \frac{dx^\mu(\lambda)}{d\lambda}\frac{\partial}{\partial x^\mu} = X^{\mu'}\frac{\partial}{\partial x^{\mu'}} = \frac{dx^{\mu'}(\lambda)}{d\lambda}\frac{\partial}{\partial x^{\mu'}}. \tag{7.19}$$

Using the chain rule, we can write

$$X^{\mu'} = \frac{dx^{\mu'}[x^\mu(\lambda)]}{d\lambda} = \frac{\partial x^{\mu'}}{\partial x^\mu}\frac{dx^\mu(\lambda)}{d\lambda} = \frac{\partial x^{\mu'}}{\partial x^\mu}X^\mu. \tag{7.20}$$

This yields the transformation rules for vector components and basis vectors:

$$X^{\mu'} = \frac{\partial x^{\mu'}}{\partial x^\mu}X^\mu, \quad \frac{\partial}{\partial x^{\mu'}} = \frac{\partial x^\mu}{\partial x^{\mu'}}\frac{\partial}{\partial x^\mu}. \tag{7.21}$$

In some classical texts this is the *definition* of a vector, that is a vector is defined as an object whose components transform in this way.

We now consider the second part of (7.14), which refers to the change of the function dependent on the coordinates. We have treated tangent vectors as operators acting on functions. Analogously we can introduce the *differential* of a function as an operator acting on tangent vectors. The corresponding vector space is the *dual* or *cotangent space* $T_p^* M$, with the differential $df \in T_p^* M$ given by

$$df = \frac{\partial f}{\partial x^\mu} dx^\mu, \tag{7.22}$$

where dx^μ are the basis vectors of $T_p^* M$, which are *dual* to the coordinate basis vectors $\partial/\partial x^\mu$ of $T_p M$:

$$\langle dx^\mu, \frac{\partial}{\partial x^\nu} \rangle = \delta^\mu_\nu. \tag{7.23}$$

The change of a function f, represented by the differential df, along a curve $c(\lambda)$, represented by the tangent vector $d/d\lambda$, is then given by

$$\frac{df}{d\lambda} = \langle df, \frac{d}{d\lambda} \rangle = \langle \frac{\partial f}{\partial x^\mu} dx^\mu, \frac{dx^\nu}{d\lambda} \frac{\partial}{\partial x^\nu} \rangle = \frac{\partial f}{\partial x^\mu} \frac{dx^\nu}{d\lambda} \langle dx^\mu, \frac{\partial}{\partial x^\nu} \rangle = \frac{\partial f}{\partial x^\mu} \frac{dx^\mu}{d\lambda}. \tag{7.24}$$

Let us conclude at this point. Starting from the change of a function along a curve, we introduced tangent vectors (change of the curve) and differentials (change of the function) as operators on each other, forming the tangent space $T_p M$ and the dual or cotangent space $T_p^* M$. The advantage of this treatment lies in the existence of these objects independently from any specific coordinate systems, in the same way as points of the manifold exist independently from any specific coordinate system.

7.1.4 Metric and Curvature

From the vector space $T_p M$ of tangent vectors at p and the vector space $T_p^* M$ of dual or cotangent vectors we can construct the Cartesian product

$$\Pi_q^r = \otimes^q T_p^* M \otimes^r T_p M, \tag{7.25}$$

which is the ordered set of q dual vectors and r tangent vectors $(\omega^{(i)}, X_{(j)})$. A *tensor* T of type (q, r) is a multilinear operator acting on elements of Π_q^r:

$$T(\omega^{(i)}, X_{(j)}) = T^{\mu_1 \dots \mu_q}_{\nu_1 \dots \nu_r} \omega^{(1)}_{\mu^1} \dots \omega^{(r)}_{\mu^q} X^{\nu_1}_{(1)} \dots X^{\nu_r}_{(r)},$$

$$T = T^{\mu_1 \dots \mu_q}_{\nu_1 \dots \nu_r} \frac{\partial}{\partial x^{\mu_1}} \otimes \dots \otimes \frac{\partial}{\partial x^{\mu_q}} \otimes dx^{\nu_1} \otimes \dots \otimes dx^{\nu_r}. \tag{7.26}$$

A *Riemannian metric* g is a type $(0, 2)$ tensor which satisfies

- The metric is symmetric, that is $g(X, Y) = g(Y, X)$ for $X, Y \in T_p M$.
- The metric is positive-definite, that is $g(X, X) \geq 0$, with $g(X, X) = 0$ only for $X = 0$.

The metric is called *pseudo-Riemannian* if the second conditioned is weakened to the form

- If for a given $Y \in T_pM$ it holds $g(X, Y) = 0$ for all $X \in T_pM$, then $Y = 0$ (*non-degenerate* metric [6]).

We can write the metric as

$$g = g_{\mu\nu}dx^\mu \otimes dx^\nu, \qquad (7.27)$$

and obtain

$$
\begin{aligned}
g(X, Y) &= g_{\mu\nu}dx^\mu \otimes dx^\nu (X^\sigma \frac{\partial}{\partial x^\sigma}, Y^\rho \frac{\partial}{\partial x^\rho}) \\
&= g_{\mu\nu}X^\sigma Y^\rho \langle dx^\mu, \frac{\partial}{\partial x^\sigma}\rangle \langle dx^\nu, \frac{\partial}{\partial x^\rho}\rangle \\
&= g_{\mu\nu}X^\sigma Y^\rho \delta^\mu_\sigma \delta^\nu_\rho \\
&= g_{\mu\nu}X^\mu Y^\nu. \qquad (7.28)
\end{aligned}
$$

In the literature the metric is often introduced together with the *line element* [2, 7]. Given infinitesimal coordinate displacements dx^μ, the line element ds^2 reads $(c = 1)$

$$ds^2 = g_{\mu\nu}dx^\mu dx^\nu, \quad d\tau^2 = -g_{\mu\nu}dx^\mu dx^\nu, \qquad (7.29)$$

where τ is the proper time.[1] From this rather informal definition not involving tangent vectors or the metric as a type $(0, 2)$ tensor, one can go on to a path through space-time represented by a parametrized curve $x^\mu(\lambda)$. Determining derivatives $dx^\mu/d\lambda$, the path length along a space like curve is defined by

$$s = \int \sqrt{g_{\mu\nu}\frac{dx^\mu}{d\lambda}\frac{dx^\nu}{d\lambda}}d\lambda, \qquad (7.30)$$

and along a timelike path as

$$\tau = \int \sqrt{-g_{\mu\nu}\frac{dx^\mu}{d\lambda}\frac{dx^\nu}{d\lambda}}d\lambda. \qquad (7.31)$$

It should be emphasized, that in this definition the parameter λ does not have to correspond to the proper time τ or the path length s. Let us consider for now only timelike paths. If we perform the calculation in (7.31), we obtain the function $\tau(\lambda)$, that is, the proper time along the curve parametrized by the parameter λ. We can invert this relation, obtaining $\lambda(\tau)$, and parametrize the path by $x^\mu(\tau)$. Using this parametrization, the respective tangent vector to the curve is denoted as the *four-velocity* U^μ:

[1]We choose the signature such that $\eta_{\mu\nu} = \text{diag}(-1, 1, 1, 1)$.

$$U^\mu = \frac{dx^\mu}{d\tau}. \tag{7.32}$$

From the (informal) definition in (7.31), we immediately obtain

$$g_{\mu\nu}U^\mu U^\nu = -1. \tag{7.33}$$

Using the more formal definition including tangent vectors $X, Y \in T_p M$ and the metric as a type $(0, 2)$ tensor, the inner product

$$g(X, Y) = g_{\mu\nu}X^\mu Y^\nu. \tag{7.34}$$

determines the *norm* of a vector X [8]

$$\|X\| = \pm\sqrt{|g(X, X)|}, \tag{7.35}$$

with the negative sign for time-like vectors, and the positive sign for space-like vectors.

Considering again a curve $x^\mu(\lambda)$ with parametrization λ, we have the tangent vector $(dx^\mu/d\lambda)\partial/\partial x^\mu \in T_p M$ at a point $p \in M$, and the proper time functional along the curve

$$\tau = \int \sqrt{-g_{\mu\nu}\frac{dx^\mu}{d\lambda}\frac{dx^\nu}{d\lambda}}\,d\lambda = \int \left\| g\left(\frac{dx^\mu}{d\lambda}, \frac{dx^\nu}{d\lambda}\right)\right\|\,d\lambda, \tag{7.36}$$

which does not depend on the parametrization λ.

We want to compare the informal definition in (7.29) and the definition involving tangent vectors in (7.34). Consider a curve defined in some chart by x^μ. Let us now calculate the proper time functional along the curve for an infinitesimal segment:

$$d\tau = \int_{\lambda}^{\lambda+d\lambda} \sqrt{-g_{\mu\nu}\frac{dx^\mu}{d\lambda'}\frac{dx^\nu}{d\lambda'}}\,d\lambda' \tag{7.37}$$

$$\approx \sqrt{-g_{\mu\nu}\left(\frac{dx^\mu}{d\lambda'}\bigg|_{\lambda'=\lambda}\right)\left(\frac{dx^\nu}{d\lambda'}\bigg|_{\lambda'=\lambda}\right)}\,d\lambda. \tag{7.38}$$

This yields

$$d\tau^2 = -g_{\mu\nu}\left[\frac{dx^\mu}{d\lambda}d\lambda\right]\left[\frac{dx^\nu}{d\lambda}d\lambda\right] = -g_{\mu\nu}dx^\mu dx^\nu, \tag{7.39}$$

where dx^μ now denotes the change of the coordinates corresponding to the movement along the curve $x^\mu(\lambda)$ under an infinitesimal parameter change $\lambda \to (\lambda + d\lambda)$.

These considerations might appear self-evident, but we want to clarify the notation with respect to the line element. Whereas the inner product of two vectors as given in (7.34) is defined for two vectors $X, Y \in T_p M$ at some point $p \in M$, there is no immediate notion of a curve involved. Of course one can define a curve crossing the point $p \in M$, such that X is the respective tangent vector, and then consider the line element of this curve as the inner product of X with itself, but the starting point from this point of view is the vector, not the curve. For the proper time functional as given in (7.36), however, one assumes some curve $x^\mu(\lambda)$ from the beginning, and then calculates the respective line elements along the curve using the norm of the well-defined tangent vectors. So from this point of view, one starts with a curve and then introduces tangent vectors as a means of calculating the proper time functional along the curve.

Up to now we considered tangent spaces $T_p M$ and the metric at one specific point p of the manifold. A *vector field* X over a manifold M is a smooth assignment of vectors to each point of M. The union of all tangent spaces is denoted as the *tangent bundle* TM:

$$TM = \bigcup_{p \in M} T_p M. \tag{7.40}$$

A vector field X is then a map

$$X : M \rightarrow TM, \, p \mapsto X(p), \tag{7.41}$$

with $\chi(M)$ denoting the set of all vector fields on M. Given a coordinate system, one usually describes a vector field by means of the respective components with respect to the coordinate basis:

$$X = X^\mu(p) \left. \frac{\partial}{\partial x^\mu} \right|_p . \tag{7.42}$$

In the same way we can define the cotangent or dual bundle T^*M, or in general tensor bundles.

We now introduce the concept of an *affine connection*, which is closely related to the *parallel displacement* of a vector. A connection on the tangent bundle TM is a map $\nabla : TM \rightarrow T^*M \otimes TM$ fulfilling additivity and the product rule of derivatives [9]:

$$\nabla(X + Y) = \nabla X + \nabla Y,$$
$$\nabla(f X) = \mathrm{d}f \otimes X + f \nabla X. \tag{7.43}$$

Here X and Y are vector fields, f is some smooth function, and T^*M is the cotangent bundle on M. We define

$$\nabla_X Y = \langle X, \nabla Y \rangle, \tag{7.44}$$

where $\langle , \rangle$ denotes the pairing between TM and T^*M. It follows, that $\nabla_X Y$ is a vector field on M, which is called the *absolute differential quotient* or the *covariant derivative*. Consider a coordinate neighborhood U of a manifold M, with local coordinates x^μ. At every point $p \in U$ we can write [5, 9]

$$\nabla \frac{\partial}{\partial x^\lambda} = \Gamma^\nu_{\mu\lambda} dx^\mu \otimes \frac{\partial}{\partial x^\nu}. \tag{7.45}$$

We sometimes use the abbreviations

$$\nabla_\mu \frac{\partial}{\partial x^\lambda} = \langle \frac{\partial}{\partial x^\mu}, \nabla \frac{\partial}{\partial x^\lambda} \rangle = \Gamma^\nu_{\mu\lambda} \frac{\partial}{\partial x^\nu}. \tag{7.46}$$

With the definition

$$\omega^\nu_\lambda = \Gamma^\nu_{\mu\lambda} dx^\mu \tag{7.47}$$

this is

$$\nabla \frac{\partial}{\partial x^\lambda} = \omega^\nu_\lambda \otimes \frac{\partial}{\partial x^\nu}. \tag{7.48}$$

A smooth vector field X can be expressed as

$$X = X^\mu \frac{\partial}{\partial x^\mu}, \tag{7.49}$$

and by definition we have

$$\begin{aligned}
\nabla X &= \nabla(X^\mu \frac{\partial}{\partial x^\mu}) \\
&= dX^\mu \otimes \frac{\partial}{\partial x^\mu} + X^\mu \nabla \frac{\partial}{\partial x^\mu} \\
&= dX^\mu \otimes \frac{\partial}{\partial x^\mu} + X^\mu \Gamma^\lambda_{\nu\mu} dx^\nu \otimes \frac{\partial}{\partial x^\lambda} \\
&= \left(\frac{\partial X^\mu}{\partial x^\nu} + \Gamma^\mu_{\nu\lambda} X^\lambda \right) dx^\nu \otimes \frac{\partial}{\partial x^\mu}.
\end{aligned} \tag{7.50}$$

which is the well-known expression for the covariant derivative [2]. We thus obtain the result, that if X^μ are the components of a vector X, the covariant derivative is a tensor with the components

$$\nabla_\nu V^\mu := V^\mu_{||\nu} = V^\mu_{|\nu} + \Gamma^\mu_{\nu\lambda} V^\lambda = \frac{\partial V^\mu}{\partial x^\nu} + \Gamma^\mu_{\nu\lambda} V^\lambda, \tag{7.51}$$

where we have introduced various notations as used in [2, 5, 7]. With our previous notations, for some tangent vector $Y \in T_p M$ we then can write

$$\begin{aligned}
\nabla_Y X &= Y^\nu \nabla_\nu X \\
&= Y^\nu (\nabla_\nu X)^\mu \frac{\partial}{\partial x^\mu} \\
&= Y^\nu \left(\frac{\partial X^\mu}{\partial x^\nu} + \Gamma^\mu_{\nu\lambda} X^\lambda \right) \frac{\partial}{x^\mu}.
\end{aligned} \tag{7.52}$$

This notion can be generalized to the covariant derivative of a general tensor. Let for example

$$T = T^{\mu\nu}_{\;\;\;\lambda} \frac{\partial}{\partial x^{\mu}} \otimes \frac{\partial}{\partial x^{\nu}} \otimes \mathrm{d}x^{\lambda}, \tag{7.53}$$

be a tensor, then its covariant derivative reads

$$\nabla T = \nabla_{\rho} T^{\mu\nu}_{\;\;\;\lambda} \frac{\partial}{\partial x^{\mu}} \otimes \frac{\partial}{\partial x^{\nu}} \otimes \mathrm{d}x^{\rho} \otimes \mathrm{d}x^{\lambda}, \tag{7.54}$$

with the components

$$\nabla_{\rho} T^{\mu\nu}_{\;\;\;\lambda} = T^{\mu\nu}_{\;\;\;\lambda\;\|\rho} \frac{\partial T^{\mu\nu}_{\;\;\;\lambda}}{\partial x^{\rho}} + \Gamma^{\mu}_{\rho\kappa} T^{\kappa\nu}_{\;\;\;\lambda} + \Gamma^{\nu}_{\rho\kappa} T^{\mu\kappa}_{\;\;\;\lambda} - \Gamma^{\kappa}_{\rho\lambda} T^{\mu\nu}_{\;\;\;\kappa}. \tag{7.55}$$

We now introduce the concept of *parallel displacement*. Let $c(\lambda)$ be a curve, and $X = (\mathrm{d}x^{\mu}/\mathrm{d}\lambda)\partial/\partial x^{\mu}$ its tangent vector. We say that a vector Y is parallel transported along $c(\lambda)$, if the covariant derivative $\nabla_X Y$ vanishes everywhere on $c(t)$, that is

$$\nabla_X Y = X^{\nu} (\nabla_{\nu} Y^{\mu}) \frac{\partial}{\partial x^{\mu}} = 0, \tag{7.56}$$

which is equivalent to

$$X^{\nu} (\nabla_{\nu} Y^{\mu}) = \frac{\mathrm{d}x^{\nu}}{\mathrm{d}\lambda} \left(\frac{\partial Y^{\mu}}{\partial x^{\nu}} + \Gamma^{\mu}_{\nu\lambda} Y^{\lambda} \right) = \frac{\mathrm{d}Y^{\mu}}{\mathrm{d}\lambda} + \Gamma^{\mu}_{\nu\lambda} \frac{\mathrm{d}x^{\nu}}{\mathrm{d}\lambda} Y^{\lambda} = 0. \tag{7.57}$$

By multiplication with $\mathrm{d}\lambda$ we get the law of parallel displacement (analogously to the formulation in [2], but with a negative sign):

$$\mathrm{d}V^{\mu} = -\Gamma^{\mu}_{\nu\lambda} \mathrm{d}x^{\nu} V^{\lambda}. \tag{7.58}$$

Up to now we introduced the metric and the affine connection as two separate mathematical structures. One can demand *metric compatibility*:

$$\nabla g = 0, \tag{7.59}$$

which is equivalent to the condition that the inner product between to parallel transported vectors remains constant along the curve [5]. It can be shown [10, 11] that such a *metric connection* can be written as

$$\nabla_{\lambda} A^{\mu} = \overset{\{\}}{\nabla}_{\lambda} A^{\mu} + K^{\mu}_{\lambda\kappa} A^{\kappa}, \tag{7.60}$$

with the definition

$$\overset{\{\}}{\nabla}_{\lambda} A^{\mu} = \frac{\partial A^{\mu}}{\partial x^{\nu}} + \left\{ \begin{matrix} \mu \\ \lambda\kappa \end{matrix} \right\} A^{\kappa}. \tag{7.61}$$

Here we have introduced the *Christoffel symbols*

$$\left\{ \begin{matrix} \lambda \\ \mu\nu \end{matrix} \right\} := -\frac{1}{2} g^{\lambda\alpha} \left(\partial_\alpha g_{\mu\nu} - \partial_\mu g_{\nu\alpha} - \partial_\nu g_{\alpha\mu} \right). \tag{7.62}$$

In the classical theory of General Relativity one has a vanishing *contorsion tensor* $K^\lambda_{\mu\nu}$ in (7.60), and thus

$$\nabla_\lambda A^\mu = \overset{\{\}}{\nabla}_\lambda A^\mu = \frac{\partial A^\mu}{\partial x^\nu} + \left\{ \begin{matrix} \mu \\ \lambda\kappa \end{matrix} \right\} A^\kappa. \tag{7.63}$$

Such a torsionless and metric-compatible connection is called a *Levi-Civita* or *Christoffel connection* [5]. In some modified theories incorporating *torsion* [10, 11], one assumes asymmetric connection coefficients, which yields the *torsion tensor*

$$S^\lambda_{\mu\nu} = \Gamma^\lambda_{[\mu\nu]} = \frac{1}{2} \left(\Gamma^\lambda_{\mu\nu} - \Gamma^\lambda_{\nu\mu} \right). \tag{7.64}$$

This torsion tensor then leads to the addition contribution in the covariant derivative (7.60), called the *contorsion tensor*:

$$K^\lambda_{\mu\nu} = - \left(S^\lambda_{\mu\nu} - S^\lambda_{\mu\nu} - S^\lambda_{\nu\mu} \right). \tag{7.65}$$

We finally introduce the Riemann curvature tensor, which depends on the connection ∇. In this short introductory chapter we will mainly present the definitions and refer the reader for more information and the physical interpretation to the literature [2, 5–7]. Given vector fields X, Y, Z, the *Riemann curvature tensor* is given by

$$\mathscr{R}(X, Y)Z = \nabla_X \nabla_Y Z - \nabla_Y \nabla_X Z - \nabla_{[X,Y]} Z, \tag{7.66}$$

where $[X, Y]$ is the Lie bracket of two vector fields, which in a coordinate basis $\{\partial/\partial_\mu\}$ can be written as

$$\left(X^\mu \frac{\partial Y^\nu}{\partial x^\mu} - Y^\mu \frac{\partial X^\nu}{\partial x^\mu} \right) \frac{\partial}{\partial x^\nu}. \tag{7.67}$$

We now define the components of the curvature tensor with respect to some coordinate basis. We write

$$\mathscr{R}(X, Y)Z = \left(\mathscr{R}^\kappa_{\lambda\nu\mu} X^\mu Y^\nu Z^\lambda \right) \frac{\partial}{\partial x^\kappa}, \tag{7.68}$$

which corresponds to

$$\mathscr{R}^\kappa_{\lambda\nu\mu} = \langle dx^\kappa, \mathscr{R}(\frac{\partial}{\partial x^\mu}, \frac{\partial}{\partial x^\nu}) \frac{\partial}{\partial x^\lambda} \rangle. \tag{7.69}$$

It can be shown [7], that the components can be written in the form

$$
\begin{aligned}
\mathcal{R}^{\kappa}_{\lambda\mu\nu} Z^{\lambda} &= \left(\nabla_{\mu}\nabla_{\nu} - \nabla_{\nu}\nabla_{\mu} \right) Z^{\kappa} + \left(\Gamma^{\lambda}_{\mu\nu} - \Gamma^{\lambda}_{\nu\mu} \right) \nabla_{\lambda} Z^{\kappa} \\
&= Z^{\kappa}_{\|\nu\|\mu} - Z^{\kappa}_{\|\mu\|\nu} + \left(\Gamma^{\lambda}_{\mu\nu} - \Gamma^{\lambda}_{\nu\mu} \right) Z^{\kappa}_{\|\lambda},
\end{aligned}
\tag{7.70}
$$

where the last expression is written in a notation analogously to the one used in [2]. The components of the Riemann tensor then can be derived in a straightforward way [2, 7]:

$$
\mathcal{R}^{\kappa}_{\lambda\mu\nu} = \frac{\partial \Gamma^{\kappa}_{\nu\lambda}}{\partial x^{\mu}} - \frac{\partial \Gamma^{\kappa}_{\mu\lambda}}{\partial x^{\nu}} + \Gamma^{\kappa}_{\mu\eta}\Gamma^{\eta}_{\nu\lambda} - \Gamma^{\kappa}_{\nu\eta}\Gamma^{\eta}_{\mu\lambda}.
\tag{7.71}
$$

For the Levi-Civita or Christoffel connection, one can derive the *Bianchi identities* of the curvature tensor $\mathcal{R}(X, Y)Z$ [5]:

$$
\mathcal{R}(X, Y)Z + \mathcal{R}(Z, X)Y + \mathcal{R}(Y, Z)X = 0,
$$
$$
(\nabla_X \mathcal{R})(Y, Z)V + (\nabla_Z \mathcal{R})(X, Y)V + (\nabla_Y \mathcal{R})(Z, X)V = 0.
\tag{7.72}
$$

With respect to some specific coordinate system, these identities read [7]

$$
\begin{aligned}
\mathcal{R}_{\rho[\lambda\mu\nu]} &= \mathcal{R}_{\rho\lambda\mu\nu} + \mathcal{R}_{\rho\mu\nu\lambda} + \mathcal{R}_{\rho\nu\lambda\mu} \\
&= 0, \\
\nabla_{[\lambda} \mathcal{R}_{\rho\lambda]\mu\nu} &= \nabla_{\lambda} \mathcal{R}_{\rho\lambda\mu\nu} + \nabla_{\rho} \mathcal{R}_{\lambda\lambda\mu\nu} + \nabla_{\lambda} \mathcal{R}_{\lambda\rho\mu\nu} \\
&= 0,
\end{aligned}
\tag{7.73}
$$

where we have written

$$
\mathcal{R}_{\rho\lambda\mu\nu} = g_{\rho\kappa} \mathcal{R}^{\kappa}_{\lambda\mu\nu}.
\tag{7.74}
$$

The Ricci tensor follows from the curvature tensor by contraction [5]:

$$
\mathcal{R}(X, Y) = \langle dx^{\nu}, \mathcal{R}(X, \frac{\partial}{\partial x^{\nu}})Y \rangle,
\tag{7.75}
$$

where we have used the same symbol $\mathcal{R}$ for both the Riemann curvature tensor and the Ricci tensor, since the context and the number of indices always gives a clear indication to which object we refer to. In terms of the components this is given by

$$
\mathcal{R}^{\mu}_{\lambda\mu\nu} = \frac{\partial \Gamma^{\mu}_{\nu\lambda}}{\partial x^{\mu}} - \frac{\partial \Gamma^{\mu}_{\mu\lambda}}{\partial x^{\nu}} + \Gamma^{\mu}_{\mu\eta}\Gamma^{\eta}_{\nu\lambda} - \Gamma^{\mu}_{\nu\eta}\Gamma^{\eta}_{\mu\lambda}.
\tag{7.76}
$$

Recall that the connection in general is a geometrical object independently defined from the metric, and so is the Riemann curvature and the Ricci tensor. If we assume a symmetric connection (that is, torsionless) which is compatible with the metric (that is, the covariant derivative of the metric vanishes), this connection is uniquely determined by the metric, and accordingly the metric and the connection are no longer independent geometrical objects. It follows, that also in this case also the curvature and Ricci tensor are determined by the metric. It is written as

$$\mathscr{R}^{\kappa}_{\lambda\mu\nu} = \partial_{\mu} \begin{Bmatrix} \kappa \\ \nu\lambda \end{Bmatrix} - \partial_{\nu} \begin{Bmatrix} \kappa \\ \mu\lambda \end{Bmatrix} + \begin{Bmatrix} \kappa \\ \eta\mu \end{Bmatrix} \begin{Bmatrix} \eta \\ \nu\lambda \end{Bmatrix} - \begin{Bmatrix} \kappa \\ \eta\nu \end{Bmatrix} \begin{Bmatrix} \eta \\ \mu\lambda \end{Bmatrix}, \tag{7.77}$$

For the scalar curvature $\mathscr{R}$ one needs both the metric and the connection (which determines the curvature):

$$\mathscr{R} = g^{\lambda\nu}\mathscr{R}_{\lambda\nu}. \tag{7.78}$$

7.2 Pseudo-complex Differential Geometry

In the last section we gave a short introduction to selected topics from *real* differential geometry. We now turn to the *pseudo-complex* case. We will observe that a *pseudo-complex manifold* corresponds to a *real product manifold*. To obtain the full pseudo-complex structure of such manifolds, we have to *pseudo-complexify* the respective tangent and cotangent spaces.

In the following we will often use the abbreviation "pc" for pseudo-complex, as well as "ph" for pseudo-holomorphic. For the definition of a pseudo-holomorphic function consult Chap. 1.

7.2.1 Pseudo-complex Manifolds as Product Manifolds

A *pseudo-complex manifold* is defined as a topological space M with a family of charts or coordinate systems $\{(U_1, \Phi_i)\}$, where

$$\Phi_i : U_i \rightarrow \mathbb{P}^m \tag{7.79}$$

is a homeomorphism and the family of sets $\{U_i\}$ covers M. The set $\mathbb{P}^m$ as the pc-equivalent to $\mathbb{R}^m$ denotes the space of all m-tupel of pseudo-complex numbers.

If (U_i, Φ_i) and (U_j, Φ_j) are two coordinate systems and $p \in U_i \cap U_j$, then the coordinate transformation

$$\Phi_{\beta} \circ \Phi_{\alpha}^{-1} : \Phi_{\alpha}(U_{\alpha} \cap U_{\beta}) \rightarrow \Phi_{\beta}(U_{\alpha} \cap U_{\beta}) \tag{7.80}$$

is a pseudo-holomorphic function. For a given chart (U_i, Φ_i), we write for a point $p \in U_{\alpha}$

$$\Phi_i(p) = x^{\mu} = x_R^{\mu} + I x_I^{\mu} = x_|^{\mu}\sigma_+ + x^{\mu}\sigma_-. \tag{7.81}$$

We also use the notation

$$x_R = \phi_i^R(p), x_R^\mu = \left(\phi_\alpha^R(p)\right)^\mu,\tag{7.82}$$

and analogously for ϕ_i^I and $\phi_i^\pm$.

For two different pseudo-complex coordinate systems x^μ and $x^{\mu'}$, the condition of pseudo-holomorphic transformation functions reads

$$\frac{\partial x_R^\mu}{\partial x_R^{\mu'}} = \frac{\partial x_I^\mu}{\partial x_I^{\mu'}}, \ \frac{\partial x_R^\mu}{\partial x_I^{\mu'}} = \frac{\partial x_I^\mu}{\partial x_R^{\mu'}},\tag{7.83}$$

or

$$\frac{\partial x_+^\mu}{\partial x_-^{\mu'}} = 0, \ \frac{\partial x_-^\mu}{\partial x_+^{\mu'}} = 0.\tag{7.84}$$

The topological space M together with the atlas $\{(U_\alpha, (\phi_\alpha^R, \phi_\alpha^I))\}$ or $\{(U_\alpha, (\phi_\alpha^+, \phi_\alpha^-))\}$ as a real $2m$-dimensional differential manifold, and with transformation functions obeying (7.83) or (7.84), respectively, it is equivalent to a pseudo-complex manifold with pseudo-complex dimension m. In the following we will use the different notations interchangeably, speaking either of the pseudo-complex coordinates x^μ, or the real coordinate pairs x_+^μ, x_-^μ or x_R^μ, x_I^μ, respectively.

The relation (7.84) shows, that a pseudo-complex manifold can be associated with a real differentiable manifold, whose atlas displays a product structure in the sense that the coordinates $x_\pm^\mu$ transform independently. This motivates to construct a pseudo-complex manifold based on a real product manifold from the start. Let $W^\pm$ be two m-dimensional real differentiable manifolds with atlas $\{(\phi_\alpha^\pm, U_\alpha^\pm)\}$, respectively. The (real) product manifold $W^+ \times W^-$ together with the atlas $\{(U_\alpha^+ \times U_\beta^-, (\phi_\alpha^+, \phi_\beta^-))\}$ then can be associated with a pseudo-complex manifold by choosing as the coordinate functions $\Phi_{\alpha\beta} = \phi_\alpha^+(p)\sigma_+ + \phi_\beta^-(q)\sigma_-$, were $(p, q) \in U_\alpha^+ \times U_\beta^-$.

7.3 Pseudo-complex Tangent and Cotangent Spaces

7.3.1 Real Tangent and Cotangent Space of a Pseudo-complex Manifold

Consider a pseudo-complex manifold M, which can be constructed from a product manifold of two real manifolds W^+ and W^-. As stated in the last section, such a pseudo-complex manifold can be either interpreted as a real differentiable manifold with a product structure, or as a pseudo-complex manifold. The coordinate bases of the (real) tangent space $T_p M$ of M as a real differentiable manifold are given by

$$\left\{\partial_\mu^R, \partial_\mu^I\right\} \text{ or } \left\{\partial_\mu^+, \partial_\mu^-\right\}, \tag{7.85}$$

where we have used abbreviations of the type $\partial_\mu^R = \frac{\partial}{\partial x_R^\mu}$. We denote the basis $\{\partial_\mu^R, \partial_\mu^I\}$ as the *associated basis*, and the basis $\{\partial_\mu^+, \partial_\mu^-\}$ as the *product basis* [12]. The corresponding dual bases of the cotangent space $T_p^* M$ are

$$\left\{dx_R^\mu, dx_I^\mu\right\}, \quad \left\{dx_+^\mu, dx_-^\mu\right\}. \tag{7.86}$$

From $x_\pm^\mu = x_R^\mu \pm x_I^\mu$ it follows

$$\partial_\mu^\pm = \frac{1}{2}\left(\partial_\mu^R \pm \partial_\mu^I\right),$$
$$dx_\pm^\mu = dx_R^\mu \pm dx_I^\mu. \tag{7.87}$$

At the point p, the set $\{\partial_\mu^+\}$ spans the subspace $T_p M^+ \subset T_p M$ (equivalently $T_p M^- \subset T_p M$), and thus it is easy to see that $T_p M = T_p M^+ \oplus T_p M^-$ is the sum of two vector spaces. Due to the product structure of the pseudo-complex manifold, this decomposition holds globally on all charts. In case we have constructed the pseudo-complex manifold from the product manifold $W^+ \times W^-$, we can also identify $T_p M^\pm$ with $T_p W^\pm$. For the cotangent space $T_p^* M$ analogous considerations hold.

Since up to now we do not consider a pseudo-complexification of the tangent and cotangent space, every tangent vector can be associated with a curve $c(\lambda)$,

$$X = \frac{d}{d\lambda}, \quad X = \frac{dx_R^\mu}{d\lambda}\partial_\mu^R + \frac{dx_I^\mu}{d\lambda}\partial_\mu^I = \frac{dx_+^\mu}{d\lambda}\partial_\mu^+ + \frac{dx_-^\mu}{d\lambda}\partial_\mu^-, \tag{7.88}$$

and every dual vector with the differential of a function f,

$$\omega = df, \quad \omega = \frac{\partial f}{\partial x_R^\mu}dx_R^\mu + \frac{\partial f}{\partial x_I^\mu}dx_I^\mu = \frac{\partial f}{\partial x_+^\mu}dx_+^\mu + \frac{\partial f}{\partial x_-^\mu}dx_-^\mu. \tag{7.89}$$

Note that here c is a curve depending on a real parameter, $c : \mathbb{R} \to M$, and the function f maps into the real numbers $f : M \to \mathbb{R}$.

If we use different coordinate systems, we have to apply the usual transformation rules for tangent vectors and dual vectors [5]. Consider two sets of coordinates $x_\pm^\mu$ and $x_\pm^{\mu'}$, where we have interpreted M as a real differentiable manifold. The transformation rule for the components of a tangent vector X in the product basis is [5]

$$X = X_+^\mu \frac{\partial}{\partial x_\mu^+} + X_-^\mu \frac{\partial}{\partial x_\mu^-} = X_+^{\mu'} \frac{\partial}{\partial x_{\mu'}^+} + X_-^{\mu'} \frac{\partial}{\partial x_{\mu'}^-}, \quad X_\pm^\mu = \frac{\partial x_\pm^\mu}{\partial x_\pm^{\mu'}} X_\pm^{\mu'}. \tag{7.90}$$

Note that for a *coordinate* basis with respect to the product coordinates, the $\pm$-components do not mix due to the pseudo-holomorphic transformation functions

between the two different charts. Nevertheless, we can use a *non-coordinate* basis
for the tangent space $T_p M$, where the $\pm$-components are mixed.

If we use the associated basis, which is also a coordinate basis, we obtain

$$X = X_R^\mu \frac{\partial}{\partial x_\mu^R} + X_I^\mu \frac{\partial}{\partial x_\mu^I} = X_R^{\mu'} \frac{\partial}{\partial x_R^{\mu'}} + X_I^{\mu'} \frac{\partial}{\partial x_I^{\mu'}},$$

$$X_R^\mu = \frac{\partial x_R^\mu}{\partial x_R^{\mu'}} X_R^{\mu'} + \frac{\partial x_R^\mu}{\partial x_I^{\mu'}} X_I^{\mu'}, \quad X_I^\mu = \frac{\partial x_I^\mu}{\partial x_R^{\mu'}} X_R^{\mu'} + \frac{\partial x_I^\mu}{\partial x_I^{\mu'}} X_I^{\mu'}, \tag{7.91}$$

with

$$\frac{\partial x_R^\mu}{\partial x_R^{\mu'}} = \frac{\partial x_I^\mu}{\partial x_I^{\mu'}}, \quad \frac{\partial x_R^\mu}{\partial x_I^{\mu'}} = \frac{\partial x_I^\mu}{\partial x_R^{\mu'}} \tag{7.92}$$

due to pseudo-holomorphicity.

We have the following relations between the partial derivative operators:

$$\frac{\partial}{\partial x_R^\mu} = \frac{\partial x_+^\nu}{\partial x_R^\mu} \frac{\partial}{\partial x_+^\nu} + \frac{\partial x_-^\nu}{\partial x_R^\mu} \frac{\partial}{\partial x_-^\nu} = \frac{\partial}{\partial x_+^\mu} + \frac{\partial}{\partial x_-^\mu},$$

$$\frac{\partial}{\partial x_I^\mu} = \frac{\partial x_+^\nu}{\partial x_I^\mu} \frac{\partial}{\partial x_+^\nu} + \frac{\partial x_-^\nu}{\partial x_I^\mu} \frac{\partial}{\partial x_-^\nu} = \frac{\partial}{\partial x_+^\mu} - \frac{\partial}{\partial x_-^\mu}, \tag{7.93}$$

which corresponds to the relations for the different basis tangent vectors in (7.87). It
follows

$$\frac{\partial x_R^\mu}{\partial x_R^{\mu'}} = \frac{1}{2}\left(\frac{\partial}{\partial x_+^{\mu'}} + \frac{\partial}{\partial x_-^{\mu'}} \right)(x_+^\mu + x_-^\mu) = \frac{1}{2}\left(\frac{\partial x_+^\mu}{\partial x_+^{\mu'}} + \frac{\partial x_-^\mu}{\partial x_-^{\mu'}} \right) = \frac{\partial x_I^\mu}{\partial x_I^{\mu'}},$$

$$\frac{\partial x_R^\mu}{\partial x_I^{\mu'}} = \frac{1}{2}\left(\frac{\partial}{\partial x_+^{\mu'}} - \frac{\partial}{\partial x_-^{\mu'}} \right)(x_+^\mu + x_-^\mu) = \frac{1}{2}\left(\frac{\partial x_+^\mu}{\partial x_+^{\mu'}} - \frac{\partial x_-^\mu}{\partial x_-^{\mu'}} \right) = \frac{\partial x_I^\mu}{\partial x_R^{\mu'}}. \tag{7.94}$$

7.3.2 Pseudo-complexified Tangent and Cotangent Space

Up to now we considered a pseudo-complex manifold M with *real* tangent and
cotangent spaces. We obtain *pseudo-complex* tangent and cotangent spaces by the
process of pseudo-complexification, which can be performed analogously to the
complex case [5]:

$$T_p M^{\mathbb{P}} = \{V_R + I V_I | V_R, V_I \in T_p M\},$$

$$T_p^* M^{\mathbb{P}} = \{\omega^R + I \omega^I | \omega^R, \omega^I \in T_p^* M\}, \tag{7.95}$$

or equivalently

$$T_p M^{\mathbb{P}} = \{V_+ \sigma_+ + V_- \sigma_- | V_+, V_- \in T_p M\},$$
$$T_p^* M^{\mathbb{P}} = \{\omega^+ \sigma_+ + \omega^- \sigma_- | \omega^+, \omega^- \in T_p^* M\}. \tag{7.96}$$

It should be emphasized that every pseudo-complex tangent (dual) vector is build up from *two* real tangent (dual) vectors. For clarity consider some pseudo-complexified tangent vector $V \in T_p M^{\mathbb{P}}$, which we can write in terms of its basis vectors as

$$
\begin{aligned}
V &= V_R + I V_I = V_R^{\mu_R} \partial_\mu^R + V_R^{\mu_I} \partial_\mu^I + I \left(V_I^{\mu_R} \partial_\mu^R + V_I^{\mu_I} \partial_\mu^I \right) \\
&= V_+ \sigma_+ + V_- \sigma_- = \left(V_+^{\mu_+} \partial_\mu^+ + V_+^{\mu_-} \partial_\mu^- \right) + \left(V_-^{\mu_+} \partial_\mu^+ + V_-^{\mu_-} \partial_\mu^- \right) \sigma_-. \tag{7.97}
\end{aligned}
$$

In the same way we can write a pseudo-complex dual vector $\omega \in T_p^* M^{\mathbb{P}}$ in terms of the basis dual vectors.

We introduce use new bases of $T_p M^{\mathbb{P}}$ and $T_p^* M^{\mathbb{P}}$:

$$
\begin{aligned}
D_\mu &= \frac{1}{2} \left(\partial_\mu^R + I \partial_\mu^I \right) = \sigma_+ \partial_\mu^+ + \sigma_- \partial_\mu^-, \\
\overline{D}_\mu &= \frac{1}{2} \left(\partial_\mu^R - I \partial_\mu^I \right) = \sigma_- \partial_\mu^+ + \sigma_+ \partial_\mu^-, \\
\mathrm{D} x^\mu &= \mathrm{d} x_R^\mu + I \mathrm{d} x_I^\mu = \sigma_+ \mathrm{d} x_+^\mu + \sigma_- \mathrm{d} x_-^\mu, \\
\overline{\mathrm{D}} x^\mu &:= \mathrm{d} x_R^\mu - I \mathrm{d} x_I^\mu = \sigma_- \mathrm{d} x_+^\mu + \sigma_+ \mathrm{d} x_-^\mu. \tag{7.98}
\end{aligned}
$$

We have defined pseudo-complex tangent and dual vectors by algebraically extending the corresponding real vector spaces. We now try to interpret these definitions in terms of pseudo-complex derivatives of pseudo-complex functions along pseudo-complex curves.

Define a pc-curve c as

$$c: \mathbb{P} \to M, \ \lambda_R + I \lambda_I = \lambda_+ \sigma_+ + \lambda_- \sigma_- = \lambda \mapsto c(\lambda). \tag{7.99}$$

The coordinate representation of the curve in a chart (U, Φ) we write as

$$
\begin{aligned}
x^\mu(\lambda) &= (\Phi \circ c(\lambda))^\mu \\
&= x_R^\mu(\lambda_R, \lambda_I) + I x_I^\mu(\lambda_R, \lambda_I) = x_+^\mu(\lambda_+, \lambda_-)\sigma_+ + x_-^\mu(\lambda_+, \lambda_-)\sigma_-. \tag{7.100}
\end{aligned}
$$

Define a pc-function f as

$$f: M \to \mathbb{P}, \ p \mapsto f(p) = f_R(p) + I f_I(p) = f_+(p)\sigma_+ + f_-(p)\sigma_-. \tag{7.101}$$

We define the operators

$$
\frac{D}{D\lambda} = \frac{1}{2}\left(\frac{\partial}{\partial\lambda_R} + I\frac{\partial}{\partial\lambda_I}\right) = \sigma_+\frac{\partial}{\partial\lambda_+} + \sigma_-\frac{\partial}{\partial\lambda_-},
$$
$$
\frac{D}{D\overline{\lambda}} = \frac{1}{2}\left(\frac{\partial}{\partial\lambda_R} - I\frac{\partial}{\partial\lambda_I}\right) = \sigma_-\frac{\partial}{\partial\lambda_+} + \sigma_+\frac{\partial}{\partial\lambda_-}. \tag{7.102}
$$

Note that these are up to now only operators containing partial derivatives. The first one is equivalent to the pseudo-complex derivative $\frac{D}{D\lambda}$ in case it acts on a pseudo-holomorphic function $g(\lambda)$, where $\lambda = \lambda_R + I\lambda_I$.

A vector $V \in T_pM^{\mathbb{P}}$ associated with a pseudo-complex curve c can be written as $V^\mu D_\mu + V^{\bar{\mu}}\overline{D}_\mu$, where

$$
\begin{aligned}
V^\mu &= \left.\frac{D}{D\lambda}x^\mu(t)\right|_{c^{-1}(p)} \\
&= \left.\frac{1}{2}\left(\frac{dx_R^\mu}{d\lambda_R} + \frac{dx_I^\mu}{d\lambda_I} + I\left(\frac{dx_R^\mu}{d\lambda_I} + \frac{dx_I^\mu}{d\lambda_R}\right)\right)\right|_{c^{-1}(p)} \\
&= \left.\left(\frac{dx_+^\mu}{d\lambda_+}\sigma_+ + \frac{dx_-^\mu}{d\lambda_-}\sigma_-\right)\right|_{c^{-1}(p)}, \tag{7.103}
\end{aligned}
$$

and

$$
\begin{aligned}
V^{\bar{\mu}} &= \left.\frac{D}{D\overline{\lambda}}x^\mu(\lambda)\right|_{c^{-1}(p)} \\
&= \left.\frac{1}{2}\left(\frac{dx_R^\mu}{d\lambda_R} - \frac{dx_I^\mu}{d\lambda_I} + I\left(\frac{dx_R^\mu}{d\lambda_I} - \frac{dx_I^\mu}{d\lambda_R}\right)\right)\right|_{c^{-1}(p)} \\
&= \left.\left(\frac{dx_+^\mu}{d\lambda_-}\sigma_+ + \frac{dx_-^\mu}{d\lambda_+}\sigma_-\right)\right|_{c^{-1}(p)}. \tag{7.104}
\end{aligned}
$$

Note that if we define two real curves[2] $c_R: \mathbb{R} \to M$, $c_I: \mathbb{R} \to M$ by $c_R = c(\cdot, \lambda_I^*)$ and $c_R(\lambda_R^*, \cdot)$, where $(\lambda_R^*, \lambda_I^*) = c^{-1}(p)$, this definition is completely equivalent to constructing a pseudo-complex tangent vector from two real tangent vectors, which correspond to these two real curves.

Analogously to the previous paragraph, we define for a pseudo-complex variable x the operator

---

[2] 'Real' in this context means, that the curve depends on *one real* parameter, whereas a pc-curve depends on *one pseudo-complex, that is two real* parameters. The coordinate representations of both real and pc-curves in our pseudo-complex manifold are pseudo-complex coordinates.

$$\frac{D}{Dx} = \frac{1}{2}\left(\frac{\partial}{\partial x_R} + I\frac{\partial}{\partial x_I}\right) = \sigma_+\frac{\partial}{\partial x_+} + \sigma_-\frac{\partial}{\partial x_-},$$

$$\frac{D}{D\overline{x}} = \frac{1}{2}\left(\frac{\partial}{\partial x_R} - I\frac{\partial}{\partial x_I}\right) = \sigma_-\frac{\partial}{\partial x_+} + \sigma_+\frac{\partial}{\partial x_-}. \tag{7.105}$$

Again, note that these are up to now only operators containing partial derivatives, with the first one corresponding to the pseudo-complex derivative $\frac{D}{Dx}$ in case it acts on a pseudo-holomorphic function $g(x)$, where $x = x_R + Ix_I$.

A general dual vector $\omega \in T_p^* M^{\mathbb{P}}$ associated with a pseudo-complex function f can be written as

$$\omega = \omega_\mu\, Dx^\mu + \omega_{\bar{\mu}}\, \overline{Dx}^\mu, \tag{7.106}$$

where

$$\omega_\mu = \frac{D}{Dx^\mu}f(x)\bigg|_{\phi^{-1}(p)} = \frac{1}{2}\left(\frac{\partial f_R}{\partial x_R^\mu} + \frac{\partial f_I}{\partial x_I^\mu} + I\left(\frac{\partial f_R}{\partial x_I^\mu} + \frac{\partial f_I}{\partial x_R^\mu}\right)\right)\bigg|_{\phi^{-1}(p)}$$

$$= \left(\frac{\partial f_+}{\partial x_+^\mu}\sigma_+ + \frac{\partial f_-}{\partial x_-^\mu}\sigma_-\right)\bigg|_{\phi^{-1}(p)}, \tag{7.107}$$

and

$$\omega_{\bar{\mu}} = \frac{D}{D\overline{x}^\mu}f(x)\bigg|_{\phi^{-1}(p)} = \frac{1}{2}\left(\frac{\partial f_R}{\partial x_R^\mu} - \frac{\partial f_I}{\partial x_I^\mu} - I\left(\frac{\partial f_R}{\partial x_I^\mu} + \frac{\partial f_I}{\partial x_R^\mu}\right)\right)\bigg|_{\phi^{-1}(p)}$$

$$= \left(\frac{\partial f_+}{\partial x_-^\mu}\sigma_+ + \frac{\partial f_-}{\partial x_+^\mu}\sigma_-\right)\bigg|_{\phi^{-1}(p)}. \tag{7.108}$$

Obviously this construction is equivalent to the construction of a pc-cotangent vector from two real cotangent vectors, which are associated with the two real functions f_R, $f_I: M \to \mathbb{R}$.

We define a pc-curve c to be pseudo-holomorphic (ph), if its coordinate representation $x^\mu(\lambda)$, where $\lambda \in \mathbb{P}$, fulfills the pc-Cauchy-Riemann relations with respect to the partial derivatives $\frac{\partial}{\partial \lambda_I}$, $\frac{\partial}{\partial \lambda_R}$. A pc-function f is defined to be pseudo-holomorphic, if its coordinate representation fulfills the pc-Cauchy-Riemann relations with respect to the partial derivatives $\frac{\partial}{\partial x_R^\mu}$, $\frac{\partial}{\partial x_I^\mu}$.

From the relations in the last paragraphs we immediately see, that if $V = V^\mu D_\mu + V^{\bar{\mu}}\overline{D}_\mu$ is a pc-vector associated to a pseudo-holomorphic curve, then $V^{\bar{\mu}} = 0$ for all μ. Analogously, if a pc-dual vector ω is associated with a pseudo-holomorphic function, then $\omega^{\bar{\mu}} = 0$ for all μ. In this case it holds

$$X^\mu = \frac{Dx^\mu(t)}{D\lambda} = \frac{dx_R^\mu}{d\lambda_R} + I\frac{dx_I^\mu}{d\lambda_R} = \frac{dx_+}{d\lambda_+}\sigma_+ + \frac{dx_-}{d\lambda_-}\sigma_-, \tag{7.109}$$

where we have used the pc-Cauchy-Riemann relations, and have written $\frac{D}{D\lambda}$ instead of $\frac{D}{D\lambda}$, since the operator now acts on a pseudo-holomorphic function and thus can be interpreted as the total pc-derivative.

Analogously we get for a pc-dual vector $\omega \in T_p^* M$ associated with a pseudo-holomorphic function f the components

$$\omega_\mu = \frac{Df(x)}{Dx^\mu} = \frac{\partial f_R}{\partial x_R^\mu} + I\frac{\partial f_R}{\partial x_I^\mu} = \frac{\partial f_+}{\partial x_+}\sigma_+ + \frac{\partial f_-}{\partial x_-}\sigma_-, \qquad (7.110)$$

where again the pc-Cauchy-Riemann relations have been used, and we used intentionally $\frac{D}{Dx}$ rather than $\frac{D}{Dx}$. It follows that for a pseudo-complex tangent vector X associated with a ph-curve c, and a pseudo-complex cotangent vector ω associated with a ph-function f, we get

$$X(f) = \langle \omega, X \rangle = X^\mu \omega_\mu = \frac{Df}{Dx^\mu}\frac{Dx^\mu}{D\lambda} = \frac{Df}{D\lambda}, \qquad (7.111)$$

which is the total pc-derivative of a ph function along a ph-curve. Thus the interpretation of tangent (dual) vectors as directional derivatives of curves (differentials of functions) is a concept completely analogous to the real case, as long as we consider only ph-curves (ph-functions). In the following we will sometimes denote a pc-tangent vector associated to a ph curve c by the derivative $\dot{c}$, and the pc-cotangent vector associated to a ph function f as the pc-differential Df.

We will now show that these tangent and dual vectors corresponding to ph curves and functions form a subspace of the respective vector spaces.[3] Define a linear map $J_p : T_p M \to T_p M$ by

$$J_p(\partial_\mu^R) = \partial_\mu^I, \quad J_p(\partial_\mu^I) = \partial_\mu^R, \qquad (7.112)$$

with $J_p^2 = \text{id}$. The extension of J_p to $V = (V_R + I V_I) \in T_p M^{\mathbb{P}}$, $V_R, V_I \in T_p M$ is defined by $J_p(V) := J_p(V_R) + I J_p(V_I)$ [5]. We thus get

$$J_p(D_\mu) = I D_\mu, \quad J_p(\overline{D}_\mu) = -I\overline{D}_\mu, \qquad (7.113)$$

and

$$T_p M^{\mathbb{P}} = T_p M^x \oplus T_p M^{\bar{x}}, \qquad (7.114)$$

where

$$T_p M^x = \{V \in T_p M^{\mathbb{P}} | J_p(V) = IV\},$$
$$T_p M^{\bar{x}} = \{V \in T_p M^{\mathbb{P}} | J_p(V) = -IV\}. \qquad (7.115)$$

---

[3]The pseudo-complexified tangent and cotangent spaces are actually not vector spaces, but modules, since they are build of the ring of pseudo-complex numbers, and not over a field like the real or complex numbers. For our purpose this difference is not of importance, so for simplicity we use the term vector space.

It is obvious that the set $\{D_\mu\}$ represents a basis for $T_p M^x$, whereas the set $\{\overline{D}_\mu\}$ represents a basis for $T_p M^{\overline{x}}$. In an analogous way we obtain $T_p^* M^x$ and $T_p^* M^{\overline{x}}$.

Note that in (7.109) only the parameter λ_R occurs, and thus the *real* curve $c_R := c(\lambda_R + I\lambda_I^*)$ determines the pc-tangent vector (here $(\lambda_R^* + I\lambda_I^*) = c^{-1}(p)$ are the parameter values corresponding to the point p). Nevertheless no information is lost, since this curves entirely determines the ph-curve $c(\lambda)$. Accordingly we could have also used the real curve $c_I(\lambda_I) := c(\lambda_R^* + I\lambda_I)$. Analogously in (7.110) only the *real*[4] function f_R occurs, although we could have chosen f_I instead. If we use the $\{\sigma_+, \sigma_-\}$ basis, we need two curves $c_+(\lambda_+) := c(\lambda_+\sigma_+ + \lambda_-^*\sigma_-)$, and $c_-(\lambda_-) := c(\lambda_+^*\sigma_+ + \lambda_-\sigma_-)$, where again $(\lambda_+^*\sigma_+ + \lambda_-^*\sigma_-) = c^{-1}(p)$.

One could ask why we have introduced ph curves and ph functions, when at the end we are able to equivalently consider one real curve and one real function anyway. The reason is, that the pc-curves and pc-functions in their coordinate representation are maps from and into the pseudo-complex numbers, and so in every step we can freely decide, if we use the $\{1, I\}$ basis or the $\{\sigma_+, \sigma_-\}$ basis. If we would have started with a real curve and a real function from the beginning, there is no coordinate-independent way to switch from one basis representation to the other.

7.4 Metric

In the following we always consider only $T_p M^x \subset T_p M^{\mathbb{P}}$ and $T_p^* M^x \subset T_p^* M^{\mathbb{P}}$, that is the *pseudo-complexified tangent and cotangent space associated with ph curves and functions*.

Assume that we have a Riemannian metric g with $g_p: T_p M \times T_p M \to \mathbb{R}$ on M as a real differentiable manifold (recall that a metric g is a tensor field, and g_p is the field at point $p \in M$). We then extend this metric to $g_p: T_p M^x \times T_p M^x \to \mathbb{P}$ by

$$g_p(V_R + IV_I, W_R + IW_I) = g_p(V_R, W_R) + g_p(V_I, W_I) + I(g_p(V_R, W_I) + g_p(V_I, W_R)), \tag{7.116}$$

where $V_R, V_I, W_R, W_I \in T_p M$. Note that we could define an even more general expression by allowing a pseudo-complex value for $g_p(V_R, W_R)$, for instance. Given a chart (Φ, U) with the usual coordinate bases, we define $g_{\mu\nu}(p): U \to \mathbb{P}$ by

$$\begin{aligned}
g_{\mu\nu}(p) &= g_p(D_\mu, D_\nu) \\
&= \frac{1}{4}\left(g_p(\partial_\mu^R, \partial_\nu^R) + g_p(\partial_\mu^I, \partial_\nu^I) + I(g_p(\partial_\mu^R, \partial_\nu^I) + g_p(\partial_\mu^I, \partial_\nu^R))\right) \\
&= g_{\mu\nu}^R(p) + Ig_{\mu\nu}^I(p) \\
&= g_p(\partial_\mu^+, \partial_\nu^+)\sigma_+ + g_p(\partial_\mu^-, \partial_\nu^-)\sigma_- \\
&= g_{\mu\nu}^+\sigma_+ + g_{\mu\nu}^-\sigma_-.
\end{aligned} \tag{7.117}$$

---

[4]Here a *real* function f denotes $f: M \to \mathbb{R}$, although $f \circ \phi_\alpha^{-1}$ *depends* on a pseudo-complex coordinate.

Here we have used the linearity of the tensors with respect to $\sigma_\pm$ and $1, I$. This linearity allows us to write for instance

$$g_p(\sigma_+ \partial_\mu^+, \sigma_- \partial_\nu^-) = \sigma_+ \cdot \sigma_- \cdot g_p(\partial_\mu^+, \partial_\nu^-) = 0. \tag{7.118}$$

The expressions $\sigma_\pm$ thus act like a kind of mathematical device which keeps the $\pm$-spaces separated. This concept remains valid also for other tensors (curvature, torsion) and the connection.

On the coordinate neighborhood U we can write

$$g_p = g_{\mu\nu}(p)\, \mathrm{D}x^\mu \otimes \mathrm{D}x^\nu. \tag{7.119}$$

Let us calculate $g_p(V, V)$, where $V = \mathrm{d}/\mathrm{d}\lambda$ with $V^\mu = \frac{\mathrm{D}x^\mu}{\mathrm{D}\lambda}$ is the pc-tangent vector on a ph curve c. Using the $\{\sigma_+, \sigma_-\}$ basis we obtain

$$\begin{aligned}
g_p(V, V) &= g_{\mu\nu}^+ V^{\mu+} V^{\nu+} \sigma_+ + g_{\mu\nu}^- V^{\mu-} V^{\nu-} \sigma_- \\
&= g_{\mu\nu}^+ \frac{\mathrm{d}x_+^\mu}{\mathrm{d}\lambda_+} \frac{\mathrm{d}x_+^\nu}{\mathrm{d}\lambda_+} \sigma_+ + g_{\mu\nu}^- \frac{\mathrm{d}x_-^\mu}{\mathrm{d}\lambda_-} \frac{\mathrm{d}x_-^\nu}{\mathrm{d}\lambda_-} \sigma_-.
\end{aligned} \tag{7.120}$$

We observe that the line element $\mathrm{D}\omega^2(V, V) = g(V, V)$ in general is *not real, but pseudo-complex*.

7.5 Connection and Curvature

Consider the pseudo-complexified tangent space $T_p M^{\mathbb{P}} = T_p M^x \oplus T_p M^{\bar{x}}$ of the pseudo-complex manifold M. The pc-tangent bundle $TM^{\mathbb{P}}$ is given by the union of all tangent spaces at the different points $p \in M$:

$$TM^{\mathbb{P}} = \bigcup_{p \in M} T_p M^{\mathbb{P}}. \tag{7.121}$$

Analogously

$$TM^x = \bigcup_{p \in M} T_p M^x, \quad TM^{\bar{x}} = \bigcup_{p \in M} T_p M^{\bar{x}}, \tag{7.122}$$

and

$$TM^{\mathbb{P}} = TM^x \cup TM^{\bar{x}}. \tag{7.123}$$

As before, we only consider pseudo-holomorphic tangent and dual vectors and demand that this restriction remains valid under the covariant derivative. That is, at every point $p \in M$ we have the ph tangent and cotangent spaces $T_p M^x$ and $T_p^* M^x$.

Consider a chart (U, Φ) of the pseudo-complex manifold M. On U, we chose the basis $\{D_\mu\}$ as the local frame field. We define the pseudo-holomorphic connection ∇ by

$$\nabla D_\lambda = \omega_\lambda^\nu \otimes D_\nu, \quad \omega_\lambda^\nu = \Gamma_{\mu\lambda}^\nu \, \mathrm{D}x^\mu. \tag{7.124}$$

Note that the coefficients $\Gamma_{\mu\lambda}^\nu$ are in general pseudo-complex numbers. On U a smooth vector field V can be expressed as

$$V = V^\mu D_\mu, \tag{7.125}$$

where V^μ is a pc-function on U. We expect the following relation to hold (analogously to the real case):

$$\begin{aligned}
\nabla V &= \nabla(V^\mu D_\mu) \\
&= \mathrm{D}V^\mu \otimes D_\mu + V^\mu \nabla D\mu \\
&= \mathrm{D}V^\mu \otimes D_\mu + V^\mu \Gamma_{\nu\mu}^\lambda \, \mathrm{D}x^\nu \otimes D_\lambda \\
&= \left(\frac{\mathrm{D}V^\mu}{\mathrm{D}x^\nu} + V^\lambda \Gamma_{\nu\lambda}^\mu \right) \mathrm{D}x^\nu \otimes D_\mu.
\end{aligned} \tag{7.126}$$

Here we have used that the pc-differential of a ph function can be written in terms of the coordinate basis using the pc-derivative:

$$\mathrm{D}V^\mu = \frac{\mathrm{D}V^\mu}{\mathrm{D}x^\nu} \, \mathrm{D}x^\nu. \tag{7.127}$$

Relation (7.126) is true, as long as the vector field V is pseudo-holomorphic in its dependence on the coordinate, that is, $\frac{\partial V^{\mu\pm}}{\partial x_\mp^\nu} = 0$. In this case, we can easily show that (7.126) is valid by using the $\sigma_\pm$-representation of the pseudo-complex tangent fields, dual fields and components:

$$\begin{aligned}
\nabla V &= \nabla \left((V_+^\mu \sigma_+ + V_-^\mu \sigma_-)(\partial_\mu^+ \sigma_+ + \partial_\mu^- \sigma_-) \right) \\
&= \nabla(V_+^\mu \partial_\mu^+)\sigma_+ + \nabla(V_-^\mu \partial_\mu^-)\sigma_- \\
&= \left[\left(\frac{\partial V_+^\mu}{\partial x_+^\nu} + V_+^\lambda (\Gamma^+)_{\nu\lambda}^\mu \right) \mathrm{d}x_+^\nu \otimes \partial_\mu^+ \right] \sigma_+ \\
&\quad + \left[\left(\frac{\partial V_-^\mu}{\partial x_-^\nu} + V_-^\lambda (\Gamma^-)_{\nu\lambda}^\mu \right) \mathrm{d}x_-^\nu \otimes \partial_\mu^- \right] \sigma_-.
\end{aligned} \tag{7.128}$$

That is, we obtain a pseudo-complex covariant derivative, which can be decomposed into a real covariant derivative in the $\sigma_\pm$-sectors, with connection coefficients $(\Gamma^\pm)_{\nu\lambda}^\mu$. It should be emphasized that here the vector field V is pseudo-holomorphic in two ways: On the one hand it holds $V \in TM^\times$, on the other hand for some chart (U, ϕ) the components of V on U are pseudo-holomorphic functions.

The generalization of the curvature tensor, Ricci tensor and scalar curvature to the corresponding pseudo-complex expressions is obtained analogously to the procedure described for the metric. In case we restrict our considerations to the pseudo-holomorphic tangent and cotangent spaces, one only has to perform the replacements

$$\partial_\mu \to D_\mu, \quad dx^\mu \to Dx^\mu, \tag{7.129}$$

and write the components as pseudo-complex numbers. Using the $\{\sigma_+, \sigma_-\}$-basis, these components can be written completely analogous to the real case:

$$
\begin{aligned}
\mathcal{R}^\kappa_{\lambda\mu\nu} &= \left(\frac{\partial(\Gamma^+)^\kappa_{\nu\lambda}}{\partial x^\mu_+} - \frac{\partial(\Gamma^+)^\kappa_{\mu\lambda}}{\partial x^\nu_+} + (\Gamma^+)^\kappa_{\mu\eta}(\Gamma^+)^\eta_{\nu\lambda} - (\Gamma^+)^\kappa_{\nu\eta}(\Gamma^+)^\eta_{\mu\lambda} \right) \sigma_+ \\
&+ \left(\frac{\partial(\Gamma^-)^\kappa_{\nu\lambda}}{\partial x^\mu_-} - \frac{\partial(\Gamma^-)^\kappa_{\mu\lambda}}{\partial x^\nu_-} + (\Gamma^-)^\kappa_{\mu\eta}(\Gamma^-)^\eta_{\nu\lambda} - (\Gamma^-)^\kappa_{\nu\eta}(\Gamma^-)^\eta_{\mu\lambda} \right) \sigma_-, \\
\mathcal{R}_{\lambda\nu} &= (\mathcal{R}^+)_{\lambda\nu}\sigma_+ + (\mathcal{R}^-)_{\lambda\nu}\sigma_-, \\
\mathcal{R} &= \left(g^{\lambda\nu}_+ (\mathcal{R}^+)_{\lambda\nu} \right) \sigma_+ + \left(g^{\lambda\nu}_- (\mathcal{R}^-)_{\lambda\nu} \right) \sigma_-. \tag{7.130}
\end{aligned}
$$

It should be emphasized that the scalar curvature is now a pseudo-complex number. By considering the σ_+- and the σ_--part separately, it can be shown that all relations (for instance the Bianchi identities) derived for the *real* curvature tensor also hold for its *pseudo-complex* equivalent.

7.6 Pseudo-complex General Relativity

In the last section we have shown, that a pseudo-complex extension of geometrical concepts in the language of differential geometry can be formulated in a straightforward way. This extension is essentially equivalent to the occurrence of two separated sectors (denoted as σ_- and σ_+). One could now go on in the usual way and define the Einstein-Hilbert action in both sectors, respectively, and perform the variation with respect to the metric, again in both sectors separately [5, 7]. This would yield two independent copies of General Relativity. In order to obtain a new theory, a new principle is needed. In the algebraic formulation of Pseudo-Complex General Relativity a modified variational principle together with a projection to the real part is proposed. Since the so-called zero divisors of pseudo-complex numbers can be considered as *generalized zeros*, it is proposed that the variation of a function has to be such a zero divisor instead of being strictly zero:

$$\delta S = \delta S_+ \sigma_+ + \delta S_- \sigma_- = \varepsilon \sigma_\pm. \tag{7.131}$$

Using this modified principle, one gains a certain freedom to add an additional term to the right hand side of Einstein's equation, interpreted as the contribution of a dark energy. See also Chap. 2 for an alternative justification, using constraints. The subsequent argument is, that this additional term necessarily has to be different from zero in order to obtain a different theory than just two copies of General Relativity. One obtains a pseudo-complex metric with $g_{\mu\nu}^+ \neq g_{\mu\nu}^-$, since the additional term occurs only in one of both sectors. This pseudo-complex metric is then mapped to a real metric.

In a first attempt to obtain a geometric formulation of pseudo-complex General Relativity, we now try to replace the algebraic mapping procedure to the real part by an approach similar to the ideas by Crumeyrolle [13]. For this purpose, we define the *real diagonal subspace* by $X_+^\mu = X_-^\mu$, or equivalently $X_I^\mu = 0$. One has to take some additional care in the construction of the pseudo-complex manifold to achieve a coordinate-invariant definition of this subspace, but we set aside these technical details and refer the reader to the literature [12–15]. If we consider a curve only in this subspace, it obviously holds

$$\frac{\widehat{dX_I^\mu}}{d\lambda} = 0, \tag{7.132}$$

where we introduced the notion that a "hat"-symbol indicates the evaluation of an expression on the real diagonal subspace. We observe that the line element on this subspace is then given by

$$\widehat{Ds^2} = \left(\widehat{g_{\mu\nu}^R} + I \widehat{g_{\mu\nu}^I} \right) \left(\frac{\widehat{dX_R^\mu}}{d\lambda} \frac{\widehat{dX_R^\nu}}{d\lambda} \right) \tag{7.133}$$

We now demand that the line element as a physical expression, which can be measured in experiments, has to obtain a *real* value in the real diagonal subspace. It follows, that we demand

$$\widehat{g_{\mu\nu}^I} = 0, \tag{7.134}$$

which is equivalent to

$$\widehat{g_{\mu\nu}^+} = \widehat{g_{\mu\nu}^-}. \tag{7.135}$$

Since the real diagonal subspace is defined by $x_+^\mu = x_-^\mu$, this means that $g_{\mu\nu}^+$ and $g_{\mu\nu}^-$ have to be the *same function* $\widehat{g_{\mu\nu}^R}$. But since the 'standard' Levi-Civita connection is entirely determined by the metric, also the curvature tensor only depends on the metric [5]. It follows, that if $g_{\mu\nu}^+$ and $g_{\mu\nu}^-$ are the same functions, also the connection and the curvature in the $\sigma_\pm$-sector coincide, respectively. Using now the standard variation principle and derive Einstein's equation by varying the Einstein-Hilbert action in both sections with respect to the metric, respectively, we obtain simply two copies of standard General Relativity, which is reduced to the same theory on the real diagonal subspace. If on the other hand we use the modified variational principle, we obtain a different metric in the $\sigma_\pm$-sector, respectively, and thus cannot fulfill the demand of a real line element.

One possible workaround would be the introduction of *torsion*, that is a connection with non-symmetric lower indices [10, 11]. For such a connection one can have the same metric in the σ_+ and σ_- sector, but a different curvature tensor, respectively. This gives rise to an additional contribution in Einstein's equation, which could be interpreted as a dark energy. Nevertheless, such a contribution would be based on the introduction of torsion, but not on the pseudo-complex structure of the manifold and a subsequent introduction of a physical space-time.

Another possibility is to introduce the condition of a real length element squared via a constraint, as was discussed in Chap. 2. This reduces the number of degrees of freedom of the metric by the number of constraints. The constraint in Chap. 2 can also be formulated in a symmetric way, namely using as a constraint the pseudo-imaginary part of the length element square:

$$\begin{aligned} 0 &= I\left(g^+_{\mu\nu}\dot{X}^\mu_+\dot{X}^\nu_+ - g^-_{\mu\nu}\dot{X}^\mu_-\dot{X}^\nu_-\right) \\ &= (\sigma_+ - \sigma_-)\left(g^+_{\mu\nu}\dot{X}^\mu_+\dot{X}^\nu_+ - g^-_{\mu\nu}\dot{X}^\mu_-\dot{X}^\nu_-\right), \end{aligned} \tag{7.136}$$

which is more symmetrical. The difficulty is to formulate constraints within a geometric differential language. Investigations are on the way.

References

1. S. Weinberg, *Gravitation and Cosmology: Principles and Applications of the General Theory of Relativity* (Wiley, 1972)
2. R. Adler, M. Bazin, M. Schiffer, *Introduction to General Relativity*, 2nd edn. (McGraw Hill, New York, 1975)
3. T. Frankel, *The Geometry of Physics*, 2nd edn. (Cambridge University Press, Cambridge, 2004)
4. J. Jost, *Geometry and Physics* (Springer, Berlin, 2009)
5. M. Nakahara, *Geometry, Topology and Physics*, 2nd edn. (Taylor and Francis, 2003)
6. S.W. Hawking, G.F.R. Ellis, *The Large Scale Structure of Space-Time* (Cambridge University Press, Cambridge, 1973)
7. S. Carroll, *Spacetime and Geometry. An Introduction to General Relativity* (Addison-Wesley, San Francisco, 2004)
8. J. Jost, *Riemannian Geometry and Geometric Analysis*, 6th edn. (Springer, Heidelberg, 2011)
9. S.S. Chern, W.H. Chen, K.S. Lam, *Lectures on Differential Geometry* (World Scientific, Singapore, 2000)
10. T. Clifton, P.G. Ferreira, A. Padilla, C. Skordis, Modified gravity and cosmology. Phys. Rep. **513**, 1 (2012)
11. R.T. Hammond, Torsion gravity. Rep. Prog. Phys. **65**, 599 (2002)
12. A. Crumeyrolle, Construction d'une nouvelle théorie unitaire en mécanique relativiste. Théorie unitaire hypercomplexe ou complexe hyperbolique. Ann. Fac. Sci. Toulouse **29**, 53 (1965)
13. A. Crumeyrolle, Variétés differentiables à coordonnées hypercomplexes. Application à une géométrisation et à une généralisation de la théorie d'Einstein-Schrödinger. Ann. Fac. Sci. Toulouse **26**, 105–137 (1962)

14. R.-L. Clerc, Résolution des équations aux connexions du cas antisymétrique de la teorie unitaire hypercomplexe. Application a un principe variationnel. Ann. de L'I.H.P. Section A **12**, 343 (1970)
15. R.-L. Clerc, Equations de champ symétriques et équations du movement sur une variété pseudoriemannienne a connexion non symétrique. Ann. de L'I.H.P. Section A **17**, 227 (1972)

Index

© Springer International Publishing Switzerland 2016

247

P.O. Hess et al., *Pseudo-Complex General Relativity*,
FIAS Interdisciplinary Science Series, DOI 10.1007/978-3-319-25061-8

Printed in the United States
By Bookmasters